Neptune's Domain

TITLES OF RELATED INTEREST

A modern introduction to international law
M. Akehurst

Political frontiers and boundaries
J. R. V. Prescott

Politics, security and development in small states
C. Clarke & A. Payne (eds)

The price of war
D. Forbes & N. Thrift

Neptune's Domain

A political geography of the sea

MARTIN IRA GLASSNER
Department of Geography, Southern Connecticut State University

Boston
UNWIN HYMAN
London   Sydney   Wellington

Unwin Hyman Inc.
955 Massachusetts Avenue, Cambridge, MA 02139, USA

Published by the Academic Division of
Unwin Hyman Ltd
15/17 Broadwick Street, London W1V 1FP, UK

Allen & Unwin (Australia) Ltd
8 Napier Street, North Sydney, NSW 2060, Australia

Allen & Unwin (New Zealand) Ltd
in association with the Port Nicholson Press Ltd
Compusales Building, 75 Ghuznee Street, Wellington 1, New Zealand

First published in 1990

Library of Congress Cataloging in Publication Data

Glassner, Martin Ira, 1932–
    Neptune's domain: a political geography of the sea/Martin Ira Glassner.
        p.   cm.
Includes bibliographical references.
ISBN 0–04–910091–2
1. Maritime law. 2. Geography, Political. I. Title.
JX4411.G55 1990   90–34537 CIP
341.4′5 – dc20

British Library Cataloguing in Publication Data

Glassner, Martin Ira, 1932–
    Neptune's domain: a political geography of the sea.
    1. Oceans. International political aspects
    I. Title
    320.1209162

ISBN 0–04–910091–2

Typeset in 10 on 12 point Bembo
by Computape (Pickering) Ltd, N. Yorkshire
and printed in Great Britain by
the University Press, Cambridge

Contents

List of figures

Preface

When Roger Jones of Unwin Hyman invited me to write this book, I had mixed feelings and took some time to ponder the matter before finally answering in the affirmative. On the one hand, such a book was badly needed. I had often wished for one to use in my own courses in political geography, marine geography and the Law of the Sea, and I would be able to make available to a wider audience portions of my work that had appeared in scholarly publications. It would also give me an opportunity to draw upon my own experience in the United States Foreign Service, as a consultant to the United Nations Development Programme in Asia and Africa, doing fieldwork in South America, as a member of a United Nations Conference on Trade and Development (UNCTAD) Group of Experts on land-locked countries, and as representative of the International Law Association and advisor to the delegation of Nepal at the Third United Nations Conference on the Law of the Sea (UNCLOS III).

On the other hand, I was already fully committed to teaching, administrative work, research and writing, and other activities, with more projects looming on the horizon. I finally decided to do the book, but getting to it took longer than I had anticipated. The delay was actually fortuitous, for the passage of time since UNCLOS III has improved our perspective, and rational analysis is easier than it was. The organization and content of the book have also benefited from additional time devoted to thinking about them.

I am indebted to several people who made most helpful comments on the first draft. I take this opportunity to thank Professor Lewis Alexander of the University of Rhode Island, Professor Roderick Ogley of the University of Sussex, Dr Robert Smith of the Department of State and an anonymous reviewer. They helped me to prepare a better book than I could have done on my own. Its faults, however, remain mine alone. I am also grateful to my colleague, Leon Yacher, who prepared all of the original maps and diagrams.

The book is organized around one unifying theme: the geographic aspects of the new Law of the Sea, as expressed primarily in the United Nations Convention on the Law of the Sea. The first two chapters provide some essential background information. Chapters 3 through 9 explain relevant provisions of the Convention. The next two chapters cover topics excluded from the Convention, and the last three chapters are more analytical and future-oriented.

After careful consideration, I decided to keep the notes to an absolute minimum, thereby sacrificing some of the aura of scholarship so highly regarded in academic circles, in favor of the readability so often lacking in contemporary textbooks. On the other hand, the bibliography is much longer than I had originally intended, and it includes all of the published sources from which I drew material for the text. It is, nevertheless, only indicative of the literally thousands of useful items available in many languages; anyone interested in pursuing further any of the topics covered in this book will have no trouble finding good material. Similarly, the illustrations are only selected examples of what is available in the professional literature, from governments and intergovernmental organizations, and from non-governmental organizations.

The political geography of the sea is a new field. Readers of this book have a rare opportunity to help to define and develop it. I welcome any suggestions along these lines and any comments on the book itself. And, of course, I would be delighted to help anyone interested in entering the field. It needs all the talent and energy it can attract.

Martin Ira Glassner
New Haven, Connecticut
April 1989

Neptune's Domain

1 Introduction

This is a book about the sea. And it is a book about our uses of the sea. But mostly it is about how we regulate our uses of the sea. Throughout the book there is material on the physical and economic geography of the sea, but our interest here is primarily in its political geography.

The political geography of the sea may be defined as that portion of the Law of the Sea that has a spatial component; that is, those elements relating to the sea's resources, legal jurisdiction over them and over areas of the sea, and the rules for using the sea. It does not, however, include those aspects of the Law of the Sea that are highly technical, legal or administrative; such matters as the structure and operation of the International Seabed Authority (see Ch. 3), the formulas for sharing the proceeds from certain marine resources and the procedures for settling disputes. It also includes large issues such as military uses of the sea and the status of the Southern Ocean that are not covered by the Law of the Sea.

The political geography of the sea is similar to the political geography of land areas, but with important differences stemming from the very different nature of the marine environment. Among these differences are the mobility of most of the "living resources" of the sea, the importance of the third dimension – depth – in the sea, and the total absence of permanent human inhabitants of the sea. These differences notwithstanding, we are able now to develop a valid and useful political geography of the sea, to understand the interrelationships between law and politics on one hand and the ways we use and function in the sea on the other.

Much of the work of developing a political geography of the sea has been done – and continues to be done – not by geographers but by lawyers and diplomats and government officials who develop and refine rules and boundaries and practices that become part of the Law of the Sea. This is the largest and oldest, and certainly one of the most complex, components of international law. (Others cover treaties, armed conflict, nationality and so on.) It sets out rules and procedures governing a large part (some 73%) of the earth's surface. We would do well to understand something of international law and the Law of the Sea before we proceed further.

On 10 December 1982 the United Nations Convention on the Law of the

Sea was signed at Montego Bay, Jamaica, by 117 States and two other entities. This was truly an historic occasion. Never before in history had so many States signed a treaty on the very first day that it was open for signature; never before had two entities that were not States – the Cook Islands and the United Nations Council for Namibia – signed a treaty of such vast scope and importance; and never before had a State (Fiji) deposited its instrument of ratification as soon as a convention was opened for signature. This was a remarkable testimony to the significance attached to this Convention by the overwhelming majority of the countries of the world. In addition, by March 1989 the Convention had received 159 signatures and 37 ratifications. It will enter into force one year after the deposit of the 60th instrument of ratification or accession. This may not occur for many years and may never occur, but already the Convention is having an enormous impact on State practice. These developments will be discussed in detail later in the book; at this point it is most important to understand that the 1982 Convention represents only one event, albeit a vital one, in the long and continuing evolution of the Law of the Sea.

The Law of the Sea

The origins of the Law of the Sea are so obscure that we may consider them lost in the legendary mists of antiquity. In classical times, the Greeks and Romans exercised various kinds of jurisdiction over their adjacent seas. Later several Italian city-states did the same. But none of the early policies produced clear and recognized sovereignty over maritime areas until Venice in the 15th century managed to enforce her claim to sovereignty over the entire Adriatic through sheer power. At about the same time, Britain claimed dominion over the North Sea, and Portugal and Spain claimed the seas off their territories in the Americas, Africa, and southern Asia. Scandinavian States attempted to control fishing, navigation, and commerce as far away as Greenland. Simultaneously, the concept of "freedom of the seas" was developing, with many navigators, sometimes backed by their sovereigns, claiming a right to sail anywhere at any time. The decisive battle between the concepts of maritime sovereignty and freedom of the seas was triggered by Spanish and Portuguese claims to all of the sea in the southern hemisphere, based on a series of papal bulls and their Treaty of Tordesillas of 1494.

The great flowering of international law took place in the 16th and 17th centuries. It was fueled by ecclesiastical disputes, the expansion of European empires, the consolidation of the European nation–state system which essentially brought an end to feudalism, the "commercial revolution" which necessitated rules for the conduct of international trade and consuls to help enforce the rules, and rapid improvements in technology which changed the nature of warfare both on land and at sea. During this period, scholars in countries such as Holland, England, Scotland, Italy and Spain, writing largely in Latin, crossed verbal swords over many issues, many of them involving the sea. During the 16th century, the concept that each State had the right to claim a territorial sea off its shores became established in international law, but its breadth continued to be a subject of intense debate.

Early in the 17th century the debate between those maritime powers advocating freedom of the seas (*mare liberum*) and those favoring the right of the State to claim as much of the sea as it can defend (*mare clausum*) reached its climax. But by this time the English had joined the Spanish and Portuguese in fighting for a *mare clausum*, leaving the Dutch, French and later others to defend the *mare liberum*. The two viewpoints were exemplified by the writings of the Dutchman Hugo de Groot (Grotius) and the Englishman John Selden. Early in the 18th century the British achieved naval supremacy and could afford to be more liberal. They became champions of the freedom of the seas and the doctrine of the *mare clausum* faded away.

Meanwhile the debate over the breadth of the territorial sea continued. From the 15th through the 18th centuries the territorial sea was considered valuable primarily as a protection for inshore fisheries against foreign fishermen. But some scholars claimed that the territorial sea should include all water within sight of land, others that it should include all water that could be defended by shore-mounted cannons, and still others argued for various fixed distances and other criteria. Both the line-of-sight and cannon-shot principles were quite imprecise, and hence difficult to standardize and enforce. The former, moreover, implied a continuous belt of territorial sea, while the latter could be applied only where cannons were actually placed on the shore. The issue became more important as other concerns were added by the coastal States to those of fishing and defense. These included customs duties, trade regulations, capturing and landing of prizes of war, tolls charged by countries controlling straits and neutrality of ships during wartime.

Gradually the Scandinavian idea of a standard territorial sea of one marine league became more widely applied. The cannon-shot rule, popularized in 1702 by Cornelius van Bynkershoek, but not originated by him, was amalgamated with the marine league concept later in the 18th century, largely because of the changing practice of States and the writings of Abbé Galiani. The Scandinavians, however, defined the marine league as being four nautical miles and thus claimed four-mile territorial seas well into the 20th century; elsewhere it was considered to be three nautical miles and most maritime countries claimed three-mile territorial seas. The Ottoman Empire and several other Mediterranean countries, however, claimed six-mile territorial seas, while other States refused to recognize a three-mile limit, claimed much broader zones of functional jurisdiction or made no claims at all. The three-mile limit, therefore, was never accepted as universal international law, and as early as 1927 the USSR initiated a new trend by claiming *territorial* waters

12 miles broad, expanding on the 12-mile fishing and customs zone that had been claimed in 1909.

THE NAUTICAL MILE

The nautical mile is widely used in air and sea navigation and the measurement of maritime boundaries. It is almost always used in Law of the Sea matters and is used uniformly in this book. It is derived from one minute (1/60 of one degree) of arc on the earth's surface. Since the earth is not a true sphere and its arcs are not uniform, a standard nautical mile approximating one minute of arc had to be defined. By international agreement it was set at 6072.12 feet or about 1.15 statute (English or land) miles or 1.852 kilometers. Wind velocity and speeds at sea or in the air are expressed in knots. A knot is one nautical mile per hour. It is thus incorrect to use the phrase "knots per hour," since a knot is a measure of both distance and time.

Besides the breadth of the territorial sea, other maritime issues vexed States and private citizens through the centuries preceding World War I. Among them were protection of fishing grounds; exercise of jurisdiction over customs, fiscal, immigration and health matters; methods of measuring the territorial sea; neutrality and security jurisdiction outside the territorial sea, and the status of bays and straits. Through the gradual development of international law, the fundamental questions were resolved, and the remaining ones seldom caused serious conflicts. But World War I brought to an end the leisurely pace of evolution of international law.

One of the most momentous results of World War I was the creation of the League of Nations. For all its faults and failures, the League represented an enormous advance in international relations. One of its major functions was the development and codification of international law, and in this field it made impressive contributions. A number of important law-making conferences were held under its auspices during its 20 years of activity. Among them, probably the most important for the development of the Law of the Sea was the Conference for the Codification of International Law, held at The Hague in the spring of 1930.

The 47 countries represented there considered an agenda which included a number of Law of the Sea items. They included: a uniform three-mile territorial sea; recognition of the claims of certain specified States to broader territorial seas; and acceptance of the principle of a maximum nine-mile contiguous zone beyond a three-mile territorial sea within which the coastal State could exercise jurisdiction in customs, sanitary and security matters. These and other proposals, such as extensive exclusive fishery zones and a "common patrimony" concept, were debated inconclusively.

Although the Hague Conference was unsuccessful in codifying the Law of the Sea, it performed a useful function in identifying and partially defining many issues which were to grow steadily in importance. The preparatory documents produced for the Conference, the replies of governments to them, the debates themselves, and the draft treaty articles produced by a committee on the basis of the debates form an extremely valuable body of material which has been mined by governments and scholars ever since. The International Law Commission in the 1950s used it to prepare for the first United Nations Conference on the Law of the Sea.

In 1930 it was clear that the major confrontation was between such distant-water fishing States as Japan and Britain wishing a narrow territorial sea and no exclusive fishing zones, and those States that wanted to protect from foreign fleets as broad an area of the adjacent sea as possible. This was a harbinger of a major shift in emphasis in the evolution of the Law of the Sea, from security and commerce to the resources of the sea. It also exemplifies a fundamental dichotomy of perspective regarding the sea. A minority of States with an extensive seafaring tradition have long looked upon the sea as friendly territory and have viewed coastlines from a maritime perspective. Most States, however, have until recently been almost exclusively land-oriented and have perceived the sea as useless or even hostile to them. These different perspectives and emphases were to influence the evolution of the Law of the Sea for two more generations.

The great sea rush

The quest for fish was among the motives that led some European States to make excessive claims to jurisdiction over parts of the sea. In general, however, this jockeying for maritime space took place among only a handful of countries which managed to keep one another in check, so that the extension of national maritime jurisdiction was relatively slow and moderate. All this changed, however, on 28 September 1945. On that day US President Harry S. Truman issued two proclamations. The first announced that the USA would regulate fisheries in those areas of the high seas contiguous to its coasts, but that these zones would continue to be high seas with no restrictions on navigation. This proclamation was quite moderate and it provoked little adverse comment. The second one, however, was explosive. It asserted US jurisdiction over the resources of its continental shelf and thereby set off the great sea rush of the 20th century.

The "Truman Proclamation," as it is now known, was motivated primarily by the wartime realization that the USA was becoming increasingly dependent upon imported petroleum to feed its oil-hungry economy and that this appetite could be fed in part by known and suspected offshore deposits. Another vital consideration at the time was the incipient dispute between the federal government and the several coastal states over ownership of offshore resources.

The operative clause of the proclamation stated that "the Government of the United States regards the natural resources of the subsoil and sea bed of the continental shelf beneath the high seas but contiguous to the coasts of the United States as appertaining to the United States, subject to its jurisdiction and control." A follow-up press release was interpreted as limiting US jurisdiction to the shelf within the 200-meter isobath (depth contour). The proclamation provided for negotiation with adjacent States over shelf boundaries "in accordance with equitable principles" and guaranteed traditional high seas rights in the superjacent water (the water covering the shelf).

Compared with other countries' subsequent maritime claims, it was quite moderate. It did, however, call to the attention of the world that there was something of great value out there besides fish, and that nothing, in international law or elsewhere, prevented a coastal State from claiming it. If the USA could simply grab the resources of over 700,000 square miles (2.4 million square kilometers) of territory, why couldn't others? They did – and soon.

Only a month later Mexico claimed all the rights of the two Truman proclamations in one proclamation by President Avila Camacho. Argentina in 1946 claimed not only its extraordinarily broad continental shelf but also its "epicontinental," or superjacent, waters. In 1947 Chile and Peru extended not only their jurisdiction but also their sovereignty over an adjacent maritime zone 200 miles wide, which incorporated nearly all of the rich fisheries of the Humboldt Current. Within a few years of the Truman Proclamations, maritime claims multiplied and escalated.

While this great sea rush was gathering force, the UN was embarking on its task of developing and codifying international law, and the Law of the Sea had a very high priority. In 1957 the General Assembly voted to convoke a full-dress Conference to develop one or more Conventions incorporating all of the Law of the Sea. Armed with four draft Conventions prepared by the International Law Commission (ILC), the United Nations Conference on the Law of the Sea convened in Geneva from February to April 1958, attended by 86 States.

The 1958 Conference dealt with essentially the same issues as the Hague Conference of 1930, but there were a number of important differences. First of all, there had been no sense of urgency in 1930 because that Conference was dominated by lawyers and tended to be doctrinal and legalistic. Only a few challengers to the established order forced occasional clarification of the real reasons underlying positions taken. A number of other factors combined to make the 1958 Conference both different from and more successful than its predecessor. Among them were the continental shelf doctrine, population pressures, advancing technology, proliferating claims to maritime space and resources, increased expectations of benefits from the sea to poor countries, the relatively greater strength of the UN, the much larger number of participants in international affairs, and, of course, the confrontation between the USA and the USSR. Not all of these factors, of course, had a positive influence, but they all did contribute to a general determination to reach agreement.

Using the ILC's drafts as bases for negotiation, backed by the resources of the UN Secretariat, bolstered by their own technical experts, and impelled by a sense of historical necessity, the delegates adopted four Conventions. These were the Conventions on the Territorial Sea and the Contiguous Zone, on the Continental Shelf, on the High Seas, and on Fishing and Conservation of the Living Resources of the High Seas.

Much of the traditional Law of the Sea was thus codified, as well as the continental shelf concept and a few other relatively new matters. However, the Conference failed to resolve some of the most difficult controversies, notably those over the breadth of the territorial sea and the establishment of exclusive fishing zones. The treaties adopted, moreover, left some ambiguities in the method of drawing straight baselines from which the territorial sea could be measured, included an open-ended definition of the continental shelf, and left a few minor questions unresolved.

In order to tie up the loose ends, the Second United Nations Conference on the Law of the Sea met in Geneva for six weeks in the spring of 1960 with 87 participants. It was short, sharp and unsuccessful. A US compromise proposal on the territorial sea – a six-mile territorial sea plus a six-mile fishing zone – was defeated by one vote.

The four Geneva Conventions on the Law of the Sea, signed in 1958, gradually acquired the requisite number of ratifications and entered into force between 30 September 1962 and 20 March 1966. Although for many States these Conventions represented a codification of the existing customary Law of the Sea (including the new continental shelf doctrine, which was adopted so quickly by the international community that one wag referred to it as "instant custom"), others were unhappy with one or more of them and none was greeted with wild enthusiasm. Their gaps and ambiguities generated new disagreements. Newly independent ex-colonies, which were not among the participants in either the 1958 or 1960 Conference, tended to ignore them. Increasingly, it became evident that, despite the careful draftsmanship of the ILC and the deft modifications by the 1958 Conference, the Conventions were simply inadequate to deal with changing circumstances.

2 UNCLOS III

In 1967 the evolution of the Law of the Sea received another explosive stimulus. On 17 August, Dr Arvid Pardo, Permanent Representative (Ambassador) of Malta to the UN, proposed to the Secretary-General the inclusion in the agenda of the 22nd Session of the General Assembly an item entitled, "Declaration and treaty concerning the reservation exclusively for peaceful purposes of the seabed and of the ocean floor, underlying the seas beyond the limits of present national jurisdiction, and the use of their resources in the interests of mankind." An accompanying memorandum suggested that the seabed and ocean floor be declared a "common heritage of mankind," and proposed the creation of an international agency to assume jurisdiction over this area "as a trustee for all countries," and control all activities therein.

Pardo's proposal set off a chain of events which are having the most profound effects in all history on the Law of the Sea. The General Assembly in December 1967 established an ad hoc Committee on the Peaceful Uses of the Seabed and Ocean Floor Beyond the Limits of National Jurisdiction, commonly known as the Seabed Committee. Its 35 members set to work but soon discovered that the task was beyond them. In 1969 the General Assembly enlarged the committee (which ultimately had 91 members) and it became essentially a preparatory committee for a new Law of the Sea Conference. Its chairman was the late Ambassador Hamilton Shirley Amerasinghe of Sri Lanka, who became President of the Third United Nations Conference on the Law of the Sea (UNCLOS III).

Why this sudden interest in the seabed and ultimately in a new Law of the Sea Conference? There are a number of reasons. First, the four Conventions adopted at Geneva in 1958, and one on land-locked States adopted at New York in 1965, left a number of serious matters unsettled and as time passed these grew more pressing. The Conventions, moreover, were far from being universally accepted and were even challenged as being insufficiently authoritative to establish binding norms.

Secondly, the pace of decolonization suddenly accelerated. Between the 1958 Geneva Conference and the end of 1967, 41 new countries joined the UN, 17 of them in 1960 alone. They developed a sense of solidarity, out of which grew the "Group of 77" poor countries which has since grown to over 120.

Figure 2.1 Polymetallic nodules. These nodules on the deck of a privately owned American research vessel have been collected from the seabed by the wire dredge basket after a television survey. On board the ship the site location is recorded and nodule samples are analyzed to determine metal content. Additional analysis is conducted at company headquarters. (Deepsea Ventures)

They shifted the emphasis in the General Assembly from political and security affairs to economic and social matters, and broke US control of the UN. They wanted "a piece of the action," both in the exploitation of new resources and the development of international law.

Third, the potential value of manganese nodules (more properly called polymetallic nodules), which were discovered on the world's first oceano-graphic research expedition by the British vessel HMS *Challenger* in 1873–6, was for the first time understood and then publicized in the early 1960s. Nodules are round or potato-shaped lumps containing at least 27 elements in varying proportions and at least 14 other constituents, also varying widely. Their chief components, however, are manganese, nickel, copper and cobalt. They litter the floors of the continental shelves and all the ocean basins and are even found in shallow waters in some freshwater lakes. They are not rocks, but are constantly being created by precipitation out of the water itself. Their potential value was estimated in trillions of dollars, and the technology has been developed to harvest and process them on a small scale.

Fourth, growing population pressure and new technology were resulting in an enormous boom in the harvest of fish, seaweed, marine mammals and other valuable "living resources" of the sea, and in the entry into marine fisheries of many new countries. Some species were being fished to the brink of extinction and something clearly had to be done.

Fifth, the superpowers were developing techniques for placing on the sea floor, even at very great depths, surveillance devices and nuclear and other weapons of mass destruction, capable of destroying not only each other but nearly everyone else as well. Their nuclear submarines were ranging the sea, posing perceived threats to many coastal countries. And their navies were reaching into waters hitherto free of great power rivalry, such as the polar seas and the Indian Ocean.

Sixth, the *Torrey Canyon* oil-spill disaster of 1967 dramatized for the world that modern science and technology were generating unexpected and danger-ous side effects. We were, in fact, seriously damaging the marine ecosystem. It was clearly a problem spanning national boundaries, yet there was no international mechanism available for dealing with it adequately.

Seventh, the International Geophysical Year (1957–8), the International Indian Ocean Expedition (1959–65), and other cooperative ventures in scientific exploration not only amassed a great deal of new information about the sea, but demonstrated more clearly than ever before how little we really knew about the marine environment. They also pointed up the importance and feasibility of international, as well as interdisciplinary, cooperation in marine scientific research.

Eighth, intensifying exploitation of the riches of the continental shelves, notably fish and petroleum, was impelling countries to stake out more and more of these riches for themselves, keeping "foreigners" as far from them as possible. By the end of 1967 about a dozen countries were already claiming some sort of jurisdiction beyond 12 nautical miles or the 200-metre isobath, and by 1970 the number had increased to 18. This was still not a very large number and it did not include any important maritime States, but it was clearly indicative of a dangerous trend.

These and many lesser developments made the work of the Seabed Committee exceedingly difficult. In an attempt to give some coherence to the deliberations and provide a suitable working document, the USA in August 1970 submitted a draft treaty which included an imaginative and potentially trendsetting proposal for a trusteeship zone, a kind of buffer zone between the 200-meter isobath – representing the outer limit of national jurisdiction on the continental shelf – and the outer edge of the continental margin. In this zone the coastal State would act in many respects as an agent of international authority (which would have jurisdiction over seabed mining), while still operating therein on its own account. This draft treaty was, unfortunately, not particularly well received.

The 25th Session of the General Assembly, in December 1970, adopted a "Declaration of Principles" that essentially restated Arvid Pardo's 1967 common heritage proposals. A second resolution authorized a new UN Conference on the Law of the Sea and specified many of its terms of reference.

It could not be comprehended at the time, and even now would be challenged by many, but these two events – the submission of the US Draft Treaty in August 1970 and the two General Assembly resolutions of December 1970 – probably were the high water mark of internationalism in the Law of the Sea debates. The more enlightened proposals, including the common heritage principle itself, were to be submerged under a rising tide of nationalism and sheer greed, all disguised, of course, by eloquent circumlocutions and citations from the holy scriptures of science, history, philosophy and law.

During the summer of 1972 the Seabed Committee adopted a list of 92 "subjects and issues" to be discussed at the forthcoming UNCLOS III. They included nearly every imaginable aspect of the Law of the Sea. While the Seabed Committee continued deliberating this agenda throughout 1972 and 1973, one of the most important issues was essentially being decided outside the UN structure entirely.

A series of regional meetings of Latin American and African States gave support to the demand of Chile, Ecuador and Peru for a 200-mile "patrimonial sea" or "economic zone" of some kind, within which the coastal State would have exclusive rights to all resources, living and non-living. These meetings – beginning at Santiago in 1952 and peaking at Santo Domingo and Yaoundé in June 1972, and Addis Ababa in May 1973 – were followed by others around the world within and without the UN system, some devoted exclusively, and others only marginally, to Law of the Sea matters. Everywhere the Latin American leaders and a few African followers kept up a steady drumfire of propaganda on behalf of a 200-mile exclusive economic zone (EEZ).

The Seabed Committee completed its work in the fall of 1973 without producing a draft treaty. UNCLOS III opened with a short organizational session at New York in December 1973 and began its substantive work at Caracas on

20 June 1974. By that time, however, the EEZ bandwagon had a lot of riders and one of the major Conference issues, except for details, had already been largely decided.

The Conference

The session at Caracas ran for a marathon 10 weeks and was followed by 14 others at Geneva and New York, plus many informal intercessional meetings around the world, before the final one at Montego Bay, Jamaica in December 1982. Many thousands of people participated in the Conference during these nine years, as delegates, Secretariat officials, observers and support personnel, and tons of documents were generated. Why was it so difficult for UNCLOS III to complete its work when UNCLOS I did its job rather neatly in only eight weeks? There is no single simple answer, as usual, but rather an intricate web of complex answers, besides basic economic and political conflicts.

Most important of all, we must recall the eight major reasons listed above

Figure 2.2 Billboard welcoming UNCLOS III participants to Caracas. The government of Venezuela expended extraordinary efforts and funds to ensure the well-being of the participants in the first working session of the Third United Nations Conference on the Law of the Sea and the success of the Conference. In recognition of this hospitality, the Conference voted to hold its final session at Caracas and the Convention was to be named after the city. After Venezuela voted against the Convention in 1982, however, the government was induced to withdraw its invitation and the Conference then accepted the invitation of Jamaica. (Martin Glassner).

for calling the Conference in the first place, and remember that this list is not exhaustive but only indicative. Then there were the characteristics of the Conference itself. Its ambitious objective was to produce a single, "generally acceptable" convention "dealing with all matters relating to the Law of the Sea," and not a series of conventions on specific subjects. The number, complexity and interrelatedness of the agenda items were staggering, if not daunting.

The number of participants – between 135 and 150 delegations per session – was triple the number that attended the Hague Conference of 1930 and nearly double the number that wrote the four Geneva Conventions of 1958. Many of the newcomers not only had no experience in creating international law but had no maritime expertise or tradition of any kind.

The "gentlemen's agreement" adopted by the General Assembly in November 1973, which provided that "there should be no voting on [substantive] matters until all efforts at consensus have been exhausted," meant in practice that there could be no final agreement on any substantive matter until there appeared to be a consensus on all. Thus it was technically impossible to identify majorities or minorities on any issue or determine which had been conclusively settled.

The dynamics of conference diplomacy were not conducive to prompt resolution of such momentous and sensitive problems as those being dealt with. Despite the legal, economic, technical and scientific nature of the topics being discussed, this was a highly political conference and every major topic automatically became political by definition. International law today, in fact, is largely being created by the political process.

UNCLOS III was organized into three "main committees," which dealt with most of the substantive questions. The First Committee, chaired by Paul Bamela Engo of Cameroon, wrestled with the Conference's single most difficult problem – establishing an international regime for the seabed and ocean floor beyond the limits of national jurisdiction. The Second Committee, chaired by Andrés Aguilar of Venezuela, dealt with a multitude of traditional Law of the Sea matters, and some newer ones, such as the EEZ. The Third Committee, chaired by Alexander Yankov of Bulgaria, was responsible for three major items: marine scientific research, preservation of the marine environment, and transfer of marine technology from the rich, "developed" countries to the poor, "developing" countries.

UNCLOS III was the largest, longest and most complex international conference in history, and one of the most important. The intricacies of conference diplomacy need not concern us here, but it is useful to understand some of its main features. First of all, as the Hague Conference had been dominated by the Great Powers, and the Geneva Conferences by the cold war, the overriding theme in UNCLOS III was a rich–poor, or North–South, confrontation. It is easy to exaggerate this lineup, however. In reality, it appeared largely in the First Committee and was not entirely rigid even there. No single country,

Figure 2.3 The official logo of UNCLOS III. This symbol was used on all official publications of UNCLOS III and continues to be used on publications of the United Nations Office for Ocean Affairs and the Law of the Sea. (United Nations)

group, or interest dominated the Conference. Customary alignments were shattered or rearranged. Science and technology were used, if at all, as instruments of national policy.

Each session of the Conference had a different mood and different procedures. Nevertheless, three interlocked themes seemed to overshadow all others: a general determination to establish some kind of international regime for the management of the mineral resources of the deep seabed; resurgent nationalism, expressed primarily in the drive by coastal States to grab as much of the sea as they possibly could; and the effort of the Group of 77 to use the Law of the Sea as one more device for helping to achieve a New International Economic Order.

The Conference participants were organized into a wide range of groups, subgroups, working groups, consulting groups, etc. The basic units were the regional groups. They were the Latin American, African, Asian, and East European groups, all of which were negotiating groups, and the Western European and Others Group, which was composed of essentially everyone left over. There were numerous others, overlapping these and one another: The Arab Group, the Islamic Group, NATO and European Community, for example, and the issue-oriented groups, such as the Archipelagic States, the Coastal States, the Oceanic Group and the Territorialists (those advocating a 200-mile territorial sea), to name a few. The Group of 77 included nearly all the countries of Africa, Asia and Latin America, including the Caribbean. The Evensen Group, organized by Norway's chief delegate, functioned as a kind of mini-Conference to try to break deadlocks on major issues.

Most bizarre of all, from a geographer's viewpoint, was the Group of Land-locked and Other Geographically Disadvantaged States. This group organized at Caracas and tried to arrest the juggernaut of expanding coastal

State jurisdiction in the sea. The definition of a land-locked State is clear enough; the definition of a "geographically disadvantaged State" is more difficult. Basically, it was supposed to mean any country which stands to gain little or nothing from adoption of a 200-mile economic zone. Their handicaps include such geographic features as short coastlines, narrow shelves, broad shelves, facing other countries across narrow stretches of water, isolation and numerous others suggested in various proposals from time to time. Using the criteria compiled by various people, and the map of "Potentially Zone-locked Countries" issued by the Geographer of the US Department of State, we can count not only the 30 land-locked countries, but another 105 which could be considered geographically disadvantaged! In other words, all but a handful of the countries participating in UNCLOS III could fall into this very elastic category.

In reality, however, admission to the group was, except for the land-locked themselves, largely a political, and not a geographical matter. Ultimately only 26 of the 105 or more potentially eligible disadvantaged States chose to join the Group – and were accepted – while Israel, which qualified on at least two grounds and applied for admission, was rejected on political grounds. Similarly, the Vatican (Holy See in UN terminology) withdrew from the group in the summer of 1976, purportedly because it was becoming too political.

There can be no such thing as a "perfect" or "ideal" Law of the Sea. Because many conflicting viewpoints and interests were presented at the Conference and had to be reconciled, negotiation, compromise and old-fashioned horse-trading were necessary in order to develop a "generally acceptable" Law of the Sea Convention. This process was complicated by the tendency toward "bloc thinking," in which the various groups worked out positions on issues and tried to remain solidly behind these positions. One result of all this negotiation (and the use of many techniques to influence public opinion, impress superiors back home, further individual political careers, soothe or inflate egos, and achieve other less-than-lofty goals) was to prolong the Conference. Another was to produce compromises which pleased no one entirely, but displeased no one enough to break up the Conference.

One vital underpinning of UNCLOS III was the so-called "package deal," an informal compromise worked out before the end of the Caracas session in 1974. Broadly speaking, the major maritime States accepted the concepts of a 12-mile (22 km.) territorial sea and a 200-mile (370 km.) economic zone (188 nautical miles beyond the territorial sea), while the Group of 77 agreed not to press for a broader territorial sea, agreed to the general principle of free navigation through and over straits and economic zones, and accepted the concept of compulsory settlement of disputes arising out of the uses of the seas. Many details had to be worked out in subsequent sessions, but on the whole the "package deal" permitted the Conference to proceed with the business of developing and codifying the Law of the Sea.

The Convention

Using the Seabed Committee's lengthy report with its numerous alternative texts as a basis for discussion, UNCLOS III labored for nine years, producing in succession a number of "main trends" papers in the Second Committee, a "single negotiating text," a "revised single negotiating text," an "informal composite negotiating text," a "revised informal composite negotiating text," an "informal draft Convention," a "draft Convention" and finally, the United Nations Convention on the Law of the Sea. At the Ninth Session of the Conference, in 1980, it was decided that the Tenth Session, to be held in New York early in 1981, would be the final session at which the Convention would be adopted, by consensus it was hoped. On the Saturday before the session was to open on Monday, 9 March 1981, however, the new Reagan administration in the USA announced that it had fired most of the US delegation, replaced them with a much smaller group of mostly novices in Law of the Sea matters, and intended to undertake a thorough review of the draft Convention, during which it would (in effect) prevent adoption of a Convention.

This unexpected blow to the Conference, just as it was drawing to a successful conclusion after so much hard work, nearly broke the spirit of most of the delegates and threw the Conference into a state of disarray and frustration. It staggered on. At the 11th Session in New York in the spring of 1982 the US delegation presented its objections to various articles of the draft Convention, refused to compromise on any of them, insisted on a recorded vote and was joined by only three other delegations in voting against it, all for their own reasons that did not coincide with those of the USA.

The American objections were largely ideological and focused on the seabed mining provisions. They included production limitations, mandatory transfer of technology, representation in the governing body of the International Seabed Area and others. In addition, Reagan objected to any potential role for the Palestine Liberation Organization, some provisions on marine scientific research and on a conference to be held in future to review the Convention. The USA subsequently refused to sign the Convention, refused to participate in the deliberations of the Preparatory Commission for the International Seabed Authority and for the International Tribunal for the Law of the Sea, exerted great pressure on its friends and allies not to sign the Convention, tried to organize a seabed mining regime parallel to that laid out in the Convention and declared an exclusive economic zone which is not entirely consistent with the relevant Convention provisions. These actions by the USA despite the overwhelming support of the international community for the Convention, do call into question the status of the Convention as an expression of the current Law of the Sea. No responsible international lawyer would claim that the 1982 United Nations Convention *is* the Law of the Sea. It does, however, represent a consensus of the international community on what the law ought to be.

Figure 2.4 Delegates signing the United Nations Convention on the Law of the Sea and the Final Act of UNCLOS III in Montego Bay, Jamaica, 10 December 1982. A record-breaking 119 delegations signed this Convention on the first day it was open for signature, including the Cook Islands and the United Nations Council for Namibia, neither of which is a sovereign State. Since then 40 other countries have signed the Convention and, as of 2 March 1989, 40 had ratified it of the 60 needed to bring it formally into force. Fiji, in another historic gesture, deposited its instrument of ratification with the Secretary-General of the United Nations here at Montego Bay. (Martin Glassner)

Similarly, no responsible political geographer would claim that the Law of the Sea is the same as the political geography of the sea. In the discussion that follows we shall use the United Nations Convention on the Law of the Sea (LOSC or the Convention) as a framework for examining the geographic aspects of the Law of the Sea. As indicated in Chapter 1, we shall omit some provisions, such as details of the International Seabed Authority, development and transfer of marine technology, and dispute settlement procedures. As interesting and important as they are (and a number of other provisions as well), they are not geographic. On the other hand, we shall cover such matters as military uses of the sea, regional arrangements and Antarctica that are clearly geographic but are mentioned only in passing in the LOSC or omitted altogether. We shall also point out some of the ambiguities of the Convention, some of its legislative history and some of the controversies over its provisions. We cannot discuss them in detail, however, partly because of insufficient space and partly because we are, after all, geographers and not lawyers.

3 *Maritime zones*

With one important exception and a few minor ones, the land areas of the earth are under the complete sovereignty of States and there are no categories or degrees of sovereignty. The major exception is Antarctica; the minor exceptions are a number of neutral zones, disputed territories, self-governing but not totally sovereign entities, and other anomalous situations which, taken together, occupy only a tiny fraction of the land area. In the sea, however, the situation is much more complex. Rather than a portion of the sea being "sovereign" territory and the rest not so, there are a number of zones in the sea, horizontal, vertical and functional, over which States exercise varying degrees of sovereignty or jurisdiction. Here we cannot explain all of the complexities of these maritime zones, but only outline their most important features.

The territorial sea

The United Nations Convention on the Law of the Sea has clarified and codified the regime of the territorial sea which, as we have seen, took centuries to evolve. "The sovereignty of a coastal State extends," reads Article 2, "beyond its land territory and internal waters ... to an adjacent belt of sea, described as the territorial sea. This sovereignty extends to the air space over the territorial sea as well as to its bed and subsoil." Thus, with one important exception, a State has the same degree of sovereignty over its territorial waters as it has over its land, and this sovereignty extends vertically as well, from the core of the earth to the heavens. Unlike the land, however, sovereignty over the territorial sea is limited by the traditional "right of innocent passage through the territorial sea" (Art. 17).

"Passage" is defined, as in 1958, as sailing through the territorial sea either from one point to another outside the internal waters of a coastal State or navigating to or from a State's port, roadstead or internal waters, and such "passage shall be continuous and expeditious" (Art. 18). The definition of "innocent," however, has become much more elaborate. It still includes the basic expression, "passage is innocent so long as it is not prejudicial to the peace, good order or security of the coastal State," but now there is also a list

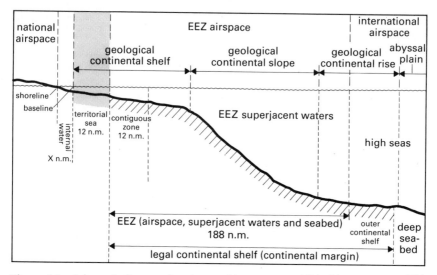

Figure 3.1   Schematic diagram showing maritime zones established by UNCLOS III. This diagram is not drawn to scale and is intended only to illustrate the terminology used in the Law of the Sea. Note that the legal continental shelf comprises the entire continental margin; that is, the submerged portions of a continent or island designated by geologists as the continental shelf, slope and rise. Note also that unlike land territory, States exercise jurisdiction for various purposes over zones that often overlap vertically. Not all States have yet claimed all of the jurisdiction to which they are entitled, while some have claimed considerably more.

of 12 activities specifically prohibited to passing vessels. Submarines are still required to navigate through the territorial sea on the surface showing their flag. Aircraft still do not have the right of innocent passage. A new article permits the coastal State to establish sea lanes and traffic separation schemes in the territorial sea for "the safety of navigation."

The centuries-long debate over the breadth of the territorial sea has essentially been resolved. According to Article 3, "Every State has the right to establish the breadth of its territorial sea up to a limit not exceeding 12 nautical miles, measured from baselines determined in accordance with this Convention." Not all States have yet exercised this right; some still claim belts of territorial waters narrower than 12 miles, and some still claim much broader belts, up to 200 miles, but the trend is clearly toward adjusting claims to comply with the new, generally accepted standard.

Internal waters and the contiguous zone

According to Article 8, all "waters on the landward side of the baseline of the territorial sea form part of the internal waters of the State," within which there is no innocent passage and the sovereignty of the State is absolute (or as absolute as any "sovereignty" can be). This includes rivers, lakes, lagoons, estuaries, ports, inter-island waters, and other water bodies, whether salt or fresh, within the baseline. The Law of the Sea does not apply in internal waters, although a State has a general obligation to respect the rights of others, to preserve the marine environment and so on. Innocent passage, however, is preserved by Article 8.2 in areas which become internal waters by being enclosed within straight baselines drawn in accordance with Article 7.

The 1958 Convention on the Territorial Sea and the Contiguous Zone provided that a State may claim a zone contiguous to the territorial sea within which it may exercise its jurisdiction over customs, fiscal, immigration and sanitary matters. The outer limit of this zone could not extend beyond 12 miles from the baseline, and the zone itself remained part of the high seas. The 1982 Convention retained the 1958 provisions intact except that the outer limit of the zone may now be 24 nautical miles from the baseline. In view of the great expansion of national jurisdiction at sea since 1958, especially the creation of the EEZ, there was some debate in UNCLOS III about whether a contiguous zone was still necessary, but since no one was strongly opposed to it, it was retained. The contiguous zone can continue to play a significant role in domestic law enforcement even against foreign vessels, but has little importance in international law.

The exclusive economic zone

We have already described the centuries-long struggle on the part of some States to extend their jurisdiction far out to sea. Most of these efforts were aimed at establishing preferential or even exclusive rights – if not outright sovereignty – over the sea's resources, primarily fish. By the time of the first substantive session of UNCLOS III at Caracas in 1974, the concept of exclusive coastal State jurisdiction over *all* resources as far as 200 miles from the baseline had already been widely accepted, and several States had already made formal claims along these lines. The campaign by Chile, Ecuador and Peru for such a zone won a few adherents first in Latin America, then West Africa, then elsewhere in Africa and Asia until the USA and a few other maritime countries conceded the point and finally everyone else went along with it. It must not be forgotten, however, that an EEZ was accepted by the major maritime powers only as part of an informal but very real "package deal," which also included a 12-mile territorial sea and transit passage through narrow straits. It was the first of many such trade-offs that ultimately became the biggest package deal of all, the Convention itself.

The wrangling in UNCLOS III over the details of the EEZ was second only to that over the seabed, and books have been written solely on this subject. Here, however, we can only outline the major features of this new regime and try to assess its consequences.

Although it does have historical antecedents, the EEZ is a new type of zone of national jurisdiction, not just an extension of existing ones such as the territorial sea and the contiguous zone. It is, in fact, unique. It is not a part of the high seas as the contiguous zone used to be, nor is it a part of the sovereign territory of a State. Within such a zone a coastal State has "sovereign rights" and "jurisdiction" and "duties," but not sovereignty. The list of rights, as might be expected, is far longer than the list of duties. Article 56 reads in part:

1   In the exclusive economic zone, the coastal State has:
    (a)   sovereign rights for the purpose of exploring and exploiting, conserving and managing the natural resources, whether living or non-living, of the waters superjacent to the seabed and of the seabed and its subsoil, and with regard to other activities for the economic exploitation and exploration of the zone, such as the production of energy from the water, currents and winds:
    (b)   jurisdiction as provided for in the relevant provisions of this Convention with regard to:

    (i)   the establishment and use of artificial islands, installations and structures;
    (ii)   marine scientific research;
    (iii) the protection and preservation of the marine environment;

    (c)   other rights and duties provided for in this Convention.

Reading this list of the rights of coastal States and their elaboration elsewhere in the Convention, it is easy to forget that not long ago Chile, Ecuador and Peru were winning sympathy for their advocacy of some kind of EEZ on the grounds that they needed to fish offshore to feed their hungry people. It is possible that these three countries will soon be leading a movement of "creeping jurisdiction;" that is, they will use various pretexts to extend their EEZs beyond 200 miles and/or convert "sovereign rights" and "jurisdiction" in the EEZ into unqualified sovereignty. Even now, however, the establishment of a 12-mile territorial sea and a 200-mile exclusive economic zone has expanded the proportion of the sea under the jurisdiction of coastal States from about 3 percent to about 36 percent. The full significance of this will not become apparent for some time, but we shall discuss some of the consequences in various contexts later in the book.

The high seas

The term "high seas" is an ancient one. It means essentially that portion of the global sea that lies beyond any kind of national jurisdiction, which currently encompasses the EEZ. The 17th-century debate among legal scholars about

Figure 3.2 The United States exclusive economic zone. This map is representational only; most of the limits shown have yet to be negotiated with opposite or adjacent States. Nevertheless, it illustrates dramatically how ownership of tiny, remote islands now gives States legal jurisdiction over vast areas of the sea. France, the United Kingdom and other rich countries also benefit in this way (as in others) from an innovation that was intended originally to benefit most of the poorer countries of the world, and that initially was vigorously opposed by the rich countries. (US Department of State)

whether the high seas was *res nullius* (belonging to no one) or *res communis* (belonging to everyone) essentially died out in the 19th century (when Britannia ruled the waves), but was revived at a lower level of passion in UNCLOS III. Though nowhere stated in so many words, the conclusion is that the high seas is community property and all members of the community have not only the right to utilize it but also the duty to protect and preserve it, to respect the rights of other States therein, and to cooperate in the maintenance of law and order.

Some of the provisions of Part VII of the 1982 LOSC are of special interest to political geographers. Among other things, they provide that

The high seas are open to all States, whether coastal or land-locked. Freedom of the high seas is exercised under the conditions laid down by this Convention and by other rules of international law. It comprises, *inter alia*, both for coastal and land-locked States:

Freedom of navigation and overflight (unrestricted) and freedom to lay submarine pipelines and cables and to construct artificial islands and installations, and freedom of fishing and scientific research (all restricted by other provisions of the Convention and of other elements of international law). (Art. 87)

"The high seas shall be reserved for peaceful purposes." (Art. 88)

"No State may validly purport to subject any part of the high seas to its sovereignty." (Art. 89)

"Every State, whether coastal or land-locked, has the right to sail ships flying its flag on the high seas." (Art. 90)

Other articles outlaw transportation of slaves, piracy, illicit traffic in narcotic drugs or psychotropic substances, and unauthorized broadcasting from the high seas, all of which, of course, represent limitations of the freedom of the seas but limitations adopted by the international community clearly for the benefit of the community. Finally, most of the 1958 Geneva Convention on Conservation of the Living Resources of the High Seas has been incorporated into Part VII of the present Convention and will be discussed in Chapter 7.

The continental shelf

All of the zones we have discussed so far (as well as archipelagic waters, which will be covered in Chapter 5) are essentially traditional or variations of traditional forms of jurisdiction over the waters of the sea and/or their resources. These are extended in the cases of internal waters, the territorial sea, and the EEZ to the core of the earth and, in the cases of internal waters, the territorial sea and the high seas, to the heavens. The remaining two zones established under the Law of the Sea are found only on the bottom of the sea and downward, thus underlying the other zones, a condition impossible on land. This is one of the features of the political geography of the sea that makes it so complex and challenging. Of these two submarine zones, that of the continental shelf has already been introduced but deserves more attention.

When studying the physical geography of the sea, we learn that the sea has drowned the fringes of all the continents and islands. Put another way, continents and islands extend out under the sea for varying distances. Continental rock (the lighter sial) is different from oceanic rock (the heavier sima) and can readily be distinguished by remote sensing, even under fairly deep layers of sediments. This submerged portion of the continent is known as

Figure 3.3 Drilling for oil on the continental shelf off the coast of Cabinda, an enclave of Angola. Nearly a third of the world's petroleum production now comes from offshore wells, mostly in such areas as Prudhoe Bay, the Gulf of Mexico, Lake Maracaibo, the North Sea and the Persian Gulf. Exploration and production fluctuate according to market prices for petroleum and its derivatives, but into the foreseeable future, this will continue to be the chief exploitable resource of the continental shelf. (United Nations/J. P. Laffont)

the continental margin, and has three parts. Working outward from shore, the first, generally falling at a relatively gentle angle, is called the continental shelf. It terminates abruptly at varying depths and distances from shore; the edge of the continental shelf, in fact, is one of the most sharply defined boundaries in nature. The margin then inclines steeply downward; this second portion is called the continental slope. Finally, the decline diminishes again and the continental rise falls away gently, to be covered by oceanic rock and sediments. There are many variations of this general pattern. In some places, such as the coasts of Namibia and Chile, coastal mountains drop sharply to the abyssal plain or even into a trench, so that there is no continental margin to speak of. In other places, such as off the Arctic coast of the USSR and the coast of Argentina, the shelf extends hundreds of miles out to sea.

These variations in the configuration of the continental margins and a general ignorance of their details led initially to the vague definition of the shelf in the 1958 Convention on the Continental Shelf, and in UNCLOS III to a prolonged, contorted and bruising battle over a new, more precise definition.

The 1958 definition was tentative, given the lack of detailed knowledge of submarine geology or of a pressing need for a precise definition at the time. It was also quite elastic. It reads in part, "the seabed and subsoil of the submarine areas adjacent to the coast but outside the area of the territorial sea, to a depth of 200 meters [the 200-meter isobath] or, beyond that limit, to where the depth of the superjacent waters admits of the exploitation of the natural resources of the said area." The 200-meter isobath had some validity at the time, though not as much as its proponents claimed; the "exploitability clause" in retrospect appears absurd.

The shelf is now defined in Article 76 of the 1982 Convention, one of the longest and most complex in the entire document. This article is, as are many provisions, the product of intense negotiations and many compromises. It reminds one of the proverbial definition of a camel as a horse designed by a committee. It begins thus:

The continental shelf of a coastal State comprises the seabed and subsoil of the submarine areas that extend beyond its territorial sea throughout the natural prolongation of its land territory to the outer edge of the continental margin, or to a distance of 200 nautical miles from the baselines from which the breadth of the territorial sea is measured where the outer edge of the continental margin does not extend up to that distance.

Not only have the 200-meter isobath and the exploitability clause been abandoned, but the geological definition has been ignored. The USSR and Argentina, for example, may claim shelf rights far beyond the actual edge of their broad shelves. Even after this principle had been accepted, defining the outer limit of the continental margin proved exceedingly difficult. The

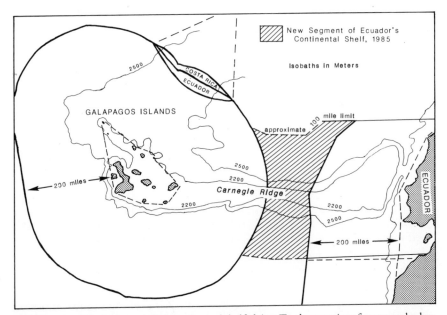

Figure 3.4 Ecuador's extended continental shelf claim. To the surprise of no one who has followed Ecuador's repeated attempts to expand its national territory farther into the Amazon Basin, the Pacific Ocean and even outer space, Ecuador has interpreted Paragraph 5 of Article 76 of the Convention in such a way as to give it exclusive jurisdiction over all of the resources of the bottom of the sea lying between her continental and insular territories. The USA and other countries have protested this action, but at present there is no effective way for them to reverse it. Such creative interpretations are likely to proliferate in coming decades. (Modified from Ecuadorian Presidential Proclamation, 19 September 1985)

subsequent paragraphs of Article 76 detail a complex formula involving the thickness of sedimentary rocks, submarine ridges, the 2,500 meter isobath, distances from the foot of the continental slope and lengths of boundary segments. The result is that a few countries will be able to claim continental shelves extending 350 nautical miles out to sea and even beyond.

"The coastal State," according to Article 77, "exercises over the continental shelf sovereign rights for the purpose of exploring it and exploiting its natural resources." In addition, it has exclusive rights to build or authorize artificial islands, installations and structures on the shelf and drilling into it "for all purposes." However, again as a result of compromises, the coastal State is obliged to "make payments or contributions in kind," again according to a complex formula, in respect of the exploitation of the non-living resources (probably mostly oil and gas) of the continental shelf beyond 200 miles. It is unlikely that this provision, Article 82, will have any practical significance for a generation or more – if ever. Meanwhile, many coastal States will be able to exploit exclusively resources which in 1967 formed part of "the common heritage of mankind."

Table 3.1 National maritime limits* at the end of 1988

Territorial sea	Contiguous zone	EEZ	Continental Shelf
107 States claim 12-mile territorial seas; other claims are:	*24 miles* Antigua/Barbuda Burma Chile Kampuchea PDR Yemen Dominica Dominican Republic Gabon Ghana India Madagascar Malta Morocco Pakistan Saint Lucia Senegal Sri Lanka Vanuatu Vietnam	74 States claim 200 miles	*200 m isobath plus exploitability* 50 states
3 miles Australia Bahamas Bahrain Belize Denmark Fed. Rep. Germany[1] Jordan Qatar Singapore Un. Arab Emirates[2] [1]Broader in parts of the North Sea [2]Sharjah only	*Others* Venezuela 3 Finland 6 Bangladesh 18 Egypt 18 The Gambia 18 Saudi Arabia 18 Sudan 18	*Fishery zone of 200 miles* Angola Antigua/Barbuda[1] Australia Bahamas Canada Denmark Dominica[1] The Gambia Fed. Rep. Germany Guyana Ireland Japan Nauru Netherlands Papua New Guinea South Africa Sweden United Kingdom [1]Also claims a 200-mile EEZ	*200 miles or outer edge of continental margin* Burma Cook Islands PDR Yemen Dominican Republic Guyana Iceland India Mauritania Mauritius New Zealand Pakistan Senegal Seychelles Sri Lanka Saint Lucia Vanuatu Vietnam
200 miles Argentina Benin Brazil Congo Ecuador El Salvador Liberia Nicaragua Panama Peru Sierra Leone Somalia Uruguay		*Others* Finland 12 Turkey 12 Malta 25 Iran 50	*Exploitability* Philippines
Others Albania 15 Angola 20 Nigeria 30 Togo 30 Syria 35 Cameroon 50 Tanzania 50			*Continental margin* Bangladesh
4 miles Finland Norway			*200 miles* Côte d'Ivoire Ghana Peru Chile[1] [1]50 miles for Sala y Gomez Island and Easter Island
6 miles Dominican Republic Greece Israel Turkey[1] [1]Aegean Sea only; 12 miles in Black and Mediterranean Seas			*200 miles or 100 miles from the 2500 m isobath* Ecuador[1] Madagascar [1]For the Galapagos Is, 100 miles from the 2500 m isobath

*in nautical miles from the applicable baseline
Source: United Nations. Office for Ocean Affairs and the Law of the Sea. 1988. *Law of the Sea Bulletin* no. 11, updated by no. 12, Dec. 1988.

The international seabed area

It may be recalled that one of the chief reasons for convoking the Third United Nations Conference on the Law of the Sea was to establish a regime for the development of the resources of "the seabed and ocean floor beyond the [present] limits of national jurisdiction," designated by the UN as "the common heritage of mankind." Very soon after Dr Pardo used these phrases, the word "present" was dropped from all repetitions of them and States around the world accelerated the extension of their maritime jurisdictions. Now the common heritage may be defined *in practice* as "what is left over after all coastal States have assumed jurisdiction over everything that could conceivably be of value for at least another generation." Considering only the EEZ and the continental shelf (as defined legally, not geologically), only about 60 percent of the seabed remains beyond these limits and is referred to simply as the Area. To be sure, this still leaves a lot of polymetallic nodules to be exploited by an international agency or under its auspices, but much of this potentially valuable resource — and others — is now reserved for the exclusive benefit of a handful of fortunate countries.

Part XI of the Convention contains 58 articles and is supplemented by two annexes with another 35 articles, and the Final Act of the Conference contains two resolutions concerning seabed mining, both of which are quite lengthy. This is a good measure of the complexity and potential importance of this pathbreaking development in international relations and in the political organization of the earth's surface. For the first time the international community has agreed to establish an agency, the International Seabed Authority, to manage and develop on its own account the resources of a huge area for the benefit of all mankind. Even if the Authority dies aborning, or fails after a long-delayed start-up, its creation on paper by and with the consent of the overwhelming majority of the States of the earth represents a signal achievement of their collective will.

The Authority, whose headquarters is to be in Kingston, Jamaica, will have an Assembly, a Council, a Secretariat and, most innovative of all, an operating arm called The Enterprise. The Authority will have the function of organizing and controlling activities in the Area, and "The Enterprise shall be the organ of the Authority which shall carry out activities in the Area directly . . . as well as the transporting, processing and marketing of minerals recovered from the Area." (Art. 170) It is to do so by means of what is called the "parallel system." A State, a private company or a consortium that wishes to harvest nodules in the Area selects two mine sites that it considers to be of roughly equal potential value, which may or may not be contiguous. The Authority then chooses one of the sites to be reserved solely for its own activities either through The Enterprise or in association with developing countries.

The entire system of mining the deep seabed and of operating the Authority and distributing the profits from mining and nearly every other aspect of

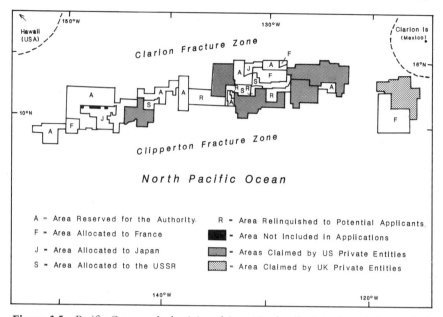

Figure 3.5 Pacific Ocean seabed mining claims. Nearly all of the formal deep seabed mining claims registered so far are in the Clarion-Clipperton Fracture Zone region between Hawaii and Mexico. This map shows the claims formally registered by the Preparatory Commission and assigned to Pioneer Investors, the areas of approximately equal potential value selected by Prepcom on behalf of the future International Seabed Authority and the claims of consortia headed by firms registered in nonsignatories such as the United States and the United Kingdom. The registered claims of India and the corresponding Authority sites are all in the Indian Ocean. There is not likely to be any large-scale commercial mining of polymetallic nodules until well into the 21st century. (Compiled from UN and US publications)

seabed operations has either been spelled out in great detail in the Convention and Final Act or assigned to the Preparatory Commission (Prepcom) to work out. Prepcom began its work in 1983, meeting twice a year alternately in New York and Kingston and at the time of writing still had a great deal of work to do.

By mid-1989, the Preparatory Commission for the International Seabed Authority and for the International Tribunal for the Law of the Sea had developed many rules and procedures for the Authority and its constitutent organs, worked on many other technical and legal matters, and laid the groundwork for the Law of the Sea Tribunal, to be located, according to the LOSC, in Hamburg, West Germany. For our purposes, Prepcom's most important accomplishments to date – and they are truly historic – have been facilitating the resolution of overlapping mining claims in the North Pacific between the USSR and Japan and between the USSR and France; formal registration of France, India, Japan and the USSR as pioneer investors, entitling them to certain advantages in seabed mining; and acceptance of designated mine sites for the Authority itself.

Table 3.2 Deep seabed mining consortia involving United States Firms

Participants	Parent company	Country of origin of parent company	Share of participation (percentage)	Main activities of parent company
USA-1 *Ocean Minerals Company* (OMCO) Mountain View, California. Formed November 1977				
Cyprus Minerals Co	Cyprus Mining Co.	United States	50	Major producers of copper, molybdenum, kaolin, gold, talc, lithium and other minerals.
Lockheed Systems Company, Inc.	Lockheed Aircraft Corporation	United States	50	Production of aircraft, missiles and spacecraft.
	Lockheed Missiles and Space Company, Inc. (subsidiary of Lockheed Aircraft Corporation)	United States		
USA-2 *Ocean Management Inc.* (OMI) New York, New York. Formed February 1975				
Inco Ltd.	International Nickel Company	Canada	25	The largest producer of nickel in the world. Also engaged in batteries and formed metal products.
AMR (Arbeitagemeinschaft Meerestechnischgewinnbare Rohstoffe)	Metallgesellschaft AG	Federal Republic of Germany	25	Mining, refining, fabricating and trading of non-ferrous metals, coal and petroleum Red Sea muds development.
	Preussag 40 Salzgitter AG			Holding company in steel making and shipbuilding.
Schlumberger Technology Corporation	Schlumberger Ltd.	Netherlands Antilles	25	Provides specialized services to drillers and producers of oil.
Deep Ocean Mining Company, Ltd. (DOMCO)	23 companies	Japan	25	Including trading, mining, and manufacturing companies and banks.
USA-3 *Ocean Mining Associates* (OMA) Gloucester Point, Virginia. Formed May 1974				
Essex Mineral Co.	United States Steel Corp.	United States	25	Steel manufacturing and fabrication.
Union Seas, Inc.	Union Minière S.A.	Belgium	25	International mining company, active in Belgium, Canada, United States, Australia.

Venture	Company	Country	%	Description
Sun Ocean Ventures	Sun Company, Inc.	United States	25	Non-operating company in oil and gas.
Samin Ocean, Inc.	Ente Nazionale Idrocarburi (ENI)	Italy	25	Italian state oil company engaged in exploration, production and marketing of oil.
USA-4 *Kennecott–Consortium* (KCON) Unincorporated. Formed January 1974				
Kennecott Corporation	Sohio	United States	40	Production and marketing of oil, coal and other minerals.
RTZ Deepsea Enterprises, Ltd.	Rio Tinto–Zinc Corporation, Ltd.	United Kingdom	12	International mining company in aluminium, copper, gold, lead and zinc.
Consolidated Gold Fields, PLC	Consgold	United Kingdom	12	International mining finance company with major interests in gold.
BP Petroleum Development, Ltd.	British Petroleum Company, Ltd.	United Kingdom	12	Major oil company with other mineral interests.
Noranda Exploration, Inc.	Noranda Mines, Ltd.	Canada	12	Mining and metallurgy of copper, lead and zinc.
Mitsubishi Group	Mitsubishi Corp. Mitsubishi Metal Corp. Mitsubishi Heavy Industries, Ltd.	Japan	12	General trading company, mining, heavy industry.

Sources: United Nations 1982. *Seabed Mineral (Resource Development)* (Sales No. E. 80. II. A. 9/Add.1.; United States 1989. NOAA *Deep Seabed Mining*.)

Other potential pioneer investors include Belgium, Canada and Italy, all members of one or more of the four officially recognized multinational consortia of firms involved in experimental seabed mining. West Germany, the UK and the USA, all of which have firms composed of their nationals participating in the multinational consortia, will not be eligible for pioneer investor status unless and until they become parties to the Convention and assume the obligations it entails.

Part XI is in some ways a microcosm of the Convention itself. It contains, for example, provisions for peaceful uses of the Area, marine scientific research, transfer of marine technology, protection of human life, accommodation of different activities in the Area and in the marine environment, archaeological and historical objects, settlement of disputes and many other matters.

One of the most unusual – and controversial – policies spelled out in the Convention is contained in Article 151, one of the longest (covering nearly three pages) in the Convention. It provides, *inter alia*, that the Authority shall control production of seabed minerals in such a way as to protect the land-based producers of the minerals from drastically falling prices that could result from the introduction into the market of vast quantities of those minerals from the seabed. This was, ostensibly at least, one of the major reasons for the refusal of the USA, UK and West Germany to sign the Convention, though it did not deter other major industrial countries, such as France, Italy and Japan, from signing.

In some ways the seabed provisions of the Convention are so numerous, complex and different from the others that they constitute virtually a separate convention. Indeed, during UNCLOS III, there was some support for producing two separate treaties, but the great majority of delegations preferred to adhere to the original mandate to produce a single, comprehensive treaty. The official view of the Reagan administration in the USA was that virtually everything in the Convention except the seabed provisions is merely a codification of customary international law and is acceptable, while the seabed portion is both new and unacceptable. This view received virtually no support, even from the USA's closest allies.

The USA argued that under the doctrine of the freedom of the high seas, anyone could mine anything on the seabed, anywhere, anytime, and the minerals would belong to whoever gets to and harvests them first. This "wild west" view was vigorously denounced by nearly all other States, which insisted that seabed minerals belonged to everyone under the common heritage principle and could only be explored and exploited under terms of the Convention, itself an indissoluble package from which no State could "pick and choose" the provisions to which it would adhere. Which view ultimately prevails remains to be seen.

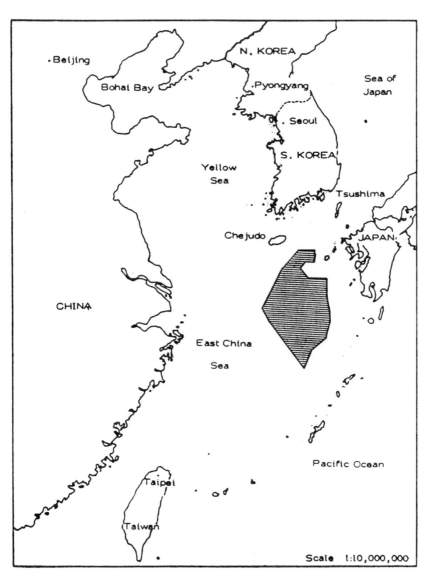

Figure 3.6 The Japan–Korea joint development zone. The continental shelf claims of Japan and South Korea overlap by 24,092 square nautical miles. The Japanese claim is based on the median lines with China and South Korea, while the South Korean claim is based on the "natural prolongation" of the peninsula toward Japan and China. In 1974 the two countries agreed that the area of overlap would be considered a joint development zone. The zone is divided into nine subzones and the agreement stipulates the terms of exploration and exploitation in some detail. Oil exploration is actively under way, but to date there have been no commercial strikes. (Park 1983, p. 131)

Maritime zones outside the treaty framework

A number of States, severally or jointly, have established zones in the sea that are not listed under any of the categories included in the Convention. Some of them are military zones of several types, and we will discuss them in Chapter 11. We will also discuss in Chapter 10 the Southern Ocean. Here we can say by way of introduction that the Southern Ocean is governed in part by the Antarctic Treaty regime established in 1959 and covering the entire region south of 60° south latitude. It is apparently excluded from application of the Law of the Sea as expressed in the 1982 Convention. Its status is quite ambiguous and is unlikely to be resolved very soon.

Because of peculiar local conditions, most commonly the inability of adjacent or opposite States to agree on maritime boundaries in areas of great resource potential, joint development zones have been established in several places. Typically, these are zones falling between uncontested boundaries in which the disputing parties agree either to share sovereignty or to ignore the sovereignty question. They agree to develop the resources of the zone jointly and to share the profits, usually as an interim measure until a boundary agreement is reached. The most important such zones at present are the one operated by Japan and South Korea in the East China Sea, and a number in the Persian Gulf. There is another shared by Malaysia and Thailand, and more may be established in the future.

It should not be surprising that such zones exist in the sea, for, as we pointed out earlier, there are similar anomalies on land and limits in the sea are much newer and less settled than those on land. It is impossible to say at this time whether these exceptions to the rules will ultimately conform and disappear as separate entities or whether they are precursors of new types of zones that themselves will become standard some day.

4 Limits in the sea

Both the traditional and the new maritime zones must be defined by boundaries or other limits if States are to exercise effectively their various types of maritime jurisdiction and if mariners and aviators are to know when they are subject to these jurisdictions. Accordingly, the definition of maritime boundaries has become much more important than it was historically, but such definition is more difficult than it is on land. Part of the difficulty arises from the fact that, unlike boundaries between States on land, maritime boundaries limit different kinds of jurisdiction that often overlap. Secondly there are few generally accepted and generally applicable principles for the drawing of limits in the sea.

Third, there are numerous islands, reefs, submarine ridges and other geological features in all the oceans – hitherto remote, unoccupied and largely neglected – whose sovereignty is disputed, and these disputes prevent the settlement of boundaries around them. Fourth, on land there are no cases of a boundary between a State and international territory (with the minor exception of the boundary between the USA and the UN headquarters site), but numerous States must have limits to their maritime territory beyond which is the International Seabed Area and/or the high seas. These limits cannot be negotiated between the parties; they can only be declared by the coastal State and accepted or rejected by the international community, generally on the basis of the Convention.

Further, there are no cultural factors to consider when drawing maritime boundaries – no linguistic or religious considerations, for example, no question of population density or splitting of families – though there are frequent questions of historic titles to parts of the sea, and historic fishing rights and similar matters to consider. There are problems of defining terms important in maritime boundary-making; "island," for example, or "natural prolongation," or "historic bay." Finally, while the stages in boundary-making on land are clear enough – definition, delimitation, demarcation and administration – in arranging maritime boundaries, States may skip the definition (description in words, usually in a treaty) and go directly to delimitation (drawing on large-scale, mutually acceptable charts). Demarcation (physically marking the boundary on the surface) is difficult or impossible, and administration (supervising the maintenance of the boundary, and policing violations of it) often difficult, especially for small or poor countries with extensive maritime zones, and even for large and rich ones. Here we can only survey some of the special problems of creating limits in the sea.

Normal and straight baselines

Since nearly all maritime zones are measured outward from a country's coastline, and since coastlines vary so greatly in their physical characteristics, there must be some standard baseline from which to measure the outer limits of the territorial sea, the contiguous zone, the EEZ and the continental shelf. Since 1958 there has been no dispute over the legal definition of a normal baseline. The 1958 definition is carried over into the 1982 Convention: "... the normal baseline for measuring the breadth of the territorial sea is the low-water line along the coast as marked on large-scale charts officially recognized by the coastal State." (Art. 5) Of course, the "low-water line" can vary considerably over time, depending on emergence or submergence of the coast, sedimentation and erosion, fluctuating tides, etc., so that the low-water line may move inward or outward by significant distances. The coastal State determines its own standard baseline (often the *mean* low-water line), and this is what is generally marked on coastal navigation charts.

There are some recognized exceptions to this general rule. For example, "in the case of islands situated on atolls or of islands having fringing reefs," says Article 6 of the LOSC, "the baseline ... is the seaward low-water line of the reef ..." Port facilities and roadsteads are also included in the territorial sea, and baselines are drawn so as to enclose them. Low-tide elevations may be used for drawing the baseline if they lie within the territorial sea, but if they lie beyond the territorial sea they have no territorial seas of their own.

Although there are still some ambiguities and uncertainties about the rules for determining normal baselines, the real problems arise with irregular coastlines. Because of its importance, Article 7, Straight Baselines, is quoted here in its entirety:

1   In localities where the coastline is deeply indented and cut into, or if there is a fringe of islands along the coast in its immediate vicinity, the method of straight baselines joining appropriate points may be employed in drawing the baseline from which the breadth of the territorial sea is measured.

2   Where because of the presence of a delta and other natural conditions the coastline is highly unstable, the appropriate points may be selected along the furthest seaward extent of the low-water line and, notwithstanding subsequent regression of the low-water line, the straight baselines shall

remain effective until changed by the coastal State in accordance with this Convention.

3 The drawing of straight baselines must not depart to any appreciable extent from the general direction of the coast, and the sea areas lying within the lines must be sufficiently closely linked to the land domain to be subject to the regime of internal waters.

4 Straight baselines shall not be drawn to and from low-tide elevations, unless lighthouses or similar installations which are permanently above sea level have been built on them or except in instances where the drawing of baselines to and from such elevations has received general international recognition.

5 Where the method of straight baselines is applicable under paragraph 1, account may be taken, in determining particular baselines, of economic interests peculiar to the region concerned, the reality and the importance of which are clearly evidenced by long usage.

6 The system of straight baselines may not be applied by a State in such a manner as to cut off the territorial sea of another State from the high seas or an exclusive economic zone.

Straight baselines were first drawn along irregular coasts by Norway, whose coast is characterized by numerous fjords and close-in islands, similar to the coasts of the Alaska panhandle and southern Chile. This was done in order to protect her inshore fisheries, simplify delimitation of her territorial sea and reduce problems of administration and enforcement of rules and laws offshore. This action was challenged by the UK but approved by the World Court in 1951. The Court's guidelines were written into the 1958 Territorial Sea Convention and retained and expanded in the 1982 Convention. Thus, Paragraphs 1 and 3 derive directly from the Anglo-Norwegian Fisheries case and are accepted as customary international law. Paragraphs 4, 5 and 6 are elaborations of these two general principles and are designed to plug some of the loopholes and reduce abuse of the concept of straight baselines. Nevertheless, some States have apparently abused the concept and claimed baselines that do not conform to these rules.

Paragraph 2, known informally as "the Bangladesh Clause," is new. It is one of many provisions of the Convention designed to benefit one or at most a very few countries, concessions necessary in the bargaining process in order to obtain wide acceptance for the Convention as a whole. (The practice of accommodating special interests, it may be noted, is quite common in legislatures around the world.)

Besides this basic article on straight baselines, there are also articles providing for the drawing of straight baselines across the mouths of rivers and enclosing bays. The first, Article 9, is simple enough: "If a river flows directly into the sea, the baseline shall be a straight line across the mouth of the river between points on the low-water line of its banks." This article does not,

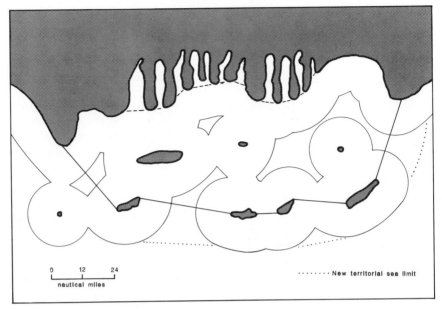

0   12   24

nautical miles

········ New territorial sea limit

Figure 4.1 Straight baselines along irregular coasts. This diagram illustrates the provisions of Article 7 of the Convention. The intent of this article is to simplify the drawing of straight baselines without unduly expanding the territorial sea. The spirit of this article is not always honored. (United Nations 1989, p. 4)

however, specify the maximum breadth of an estuary that might be subject to this article, nor does it indicate how the "points on the low-water line" are to be selected. Natually, then, States tend to interpret this article as broadly as possible so as to enclose within the baseline the maximum possible area of internal waters.

Article 10, covering "bays the coasts of which belong to a single State," is much longer and more complex. Essentially, the four operative paragraphs are designed to permit certain bays to be classified as internal waters, those that can be delimited by closing lines not more than 24 miles long, and to exclude from this classification gulfs and "mere curvature[s] of the coast." Again, despite the detailed instructions for drawing straight baselines, or closing lines, across the mouth of a bay, there are enough uncertainties so that States can interpret the letter of the article while still violating its spirit.

Also, Paragraph 6 says, "The foregoing provisions do not apply to so-called 'historic' bays, or in any case where the system of straight baselines provided for in Article 7 is applied." This opens two more loopholes, or at least leaves room for interpretation. The principle of exempting bodies of water which do not conform to the Convention's definition of "bay" but which have historically been considered as internal waters is a reasonable one. Application of the principle, however, sometimes leads to serious controversies. Perhaps

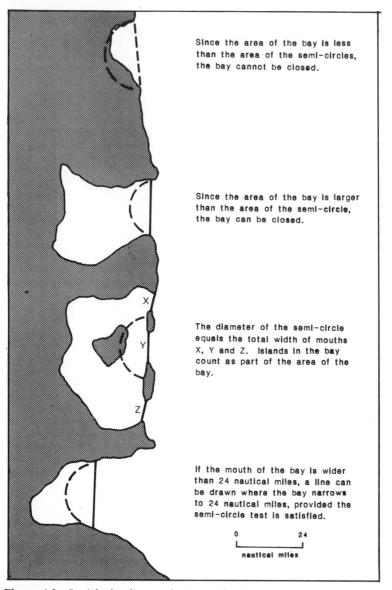

Since the area of the bay is less than the area of the semi-circles, the bay cannot be closed.

Since the area of the bay is larger than the area of the semi-circle, the bay can be closed.

The diameter of the semi-circle equals the total width of mouths X, Y and Z. Islands in the bay count as part of the area of the bay.

If the mouth of the bay is wider than 24 nautical miles, a line can be drawn where the bay narrows to 24 nautical miles, provided the semi-circle test is satisfied.

0   24

nautical miles

Figure 4.2 Straight baselines enclosing juridical bays. In the Law of the Sea, geographical definitions are often irrelevant, as in the case of "bay" vs. "gulf." The legal, or juridical, definition of a bay is complex, detailed and still subject to interpretation. This diagram illustrates some of the simpler and more common situations described in Article 10 of the Convention. There are many more – and many disagreements about them. (United Nations 1989, p. 29)

the best known are those over Hudson Bay in Canada and Peter the Great Bay on the Pacific coast of the USSR.

Boundaries between opposite or adjacent zones

When coastal State jurisdiction extended no more than three, four or even six miles out to sea, there were not many serious disputes over how to draw maritime boundaries between countries adjacent to one another on the same coast or facing one another across a body of water. Generally, either the land boundary was projected out to sea for the required distance, or a new maritime boundary was drawn perpendicular to the coast, or the median line (a line equidistant between the two coasts) was selected, or some other boundary was agreed upon, always with due regard for islands, headlands and other topographic features. Sometimes neither party considered the matter particularly important and there simply were no maritime boundaries. Now, however, the situation is very different.

There are now many more coastal countries than ever before in history. There are new, extensive zones of national jurisdiction in the sea. And there are both greater needs for developing the resources of the sea and better technology for doing so than ever before. The potential, therefore, for disagreements, disputes, and even clashes between States over maritime jurisdiction has never been greater. One of the principal reasons for convening UNCLOS III was to enable the international community to agree on rules for determining maritime jurisdiction and thus avoid the squabbles that were portended by the "cod wars" between Iceland and Britain, the "lobster war" between Brazil and France and similar episodes that punctuated the 1950s and 1960s.

At UNCLOS III, the question of delimiting the territorial sea, now legally set at up to 12 miles broad, was not particularly controversial. Delimitation has to be by agreement between the parties, and no specific guidelines or principles are provided in the Convention beyond forbidding (in Art. 15) either of two opposite States, "failing agreement between them to the contrary, to extend its territorial sea beyond the median line . . ." There is an escape clause, voiding this provision "where it is necessary by reason of historic title or other special circumstances to delimit the territorial seas of the two States in a way which is at variance therewith." There are no guidelines for drawing territorial sea boundaries between adjacent States; this must continue to be done by negotiation. Neither is there any reference to delimitation of the contiguous zone. Apparently it is simply assumed that whatever method is chosen to delimit the territorial sea will also be applied to the contiguous zone.

The question of delimitation of the continental shelf and of the exclusive economic zone is a very different matter altogether. It was so controversial at UNCLOS III, in fact, that battle lines were drawn and sides lined up by

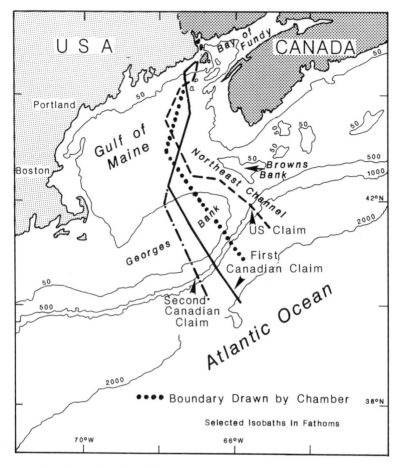

Figure 4.3 The single maritime boundary delimitation in the Gulf of Maine. The United States and Canada, impelled by both intensified fishing over and the prospect of petroleum under Georges Bank and other portions of the Gulf of Maine, agreed that there should be a single boundary in the area delimiting both their continental shelves and their EEZs, but could not agree on where the boundary should be. Canada made claims in 1977 and 1979, and the United States responded in 1977 and 1982 (to the northeast of the earlier claim). They submitted their dispute to a chamber of the International Court of Justice which rendered its decision in 1984. In essence, the chamber rejected most of the arguments on both sides, drawing a boundary that roughly bisects the disputed area. (Compiled from Canadian and American official maps)

mid-Conference, and had hardened by 1980. It was the only significant question still unresolved at the Eleventh Session in 1982. At issue was whether such delimitation should be carried out using the median line or on the basis of "equitable principles." In the end, because of the pressure to conclude the treaty, the combatants agreed to settle the issue by avoiding a settlement. Neither the median line nor equitable principles was adopted, nor was there

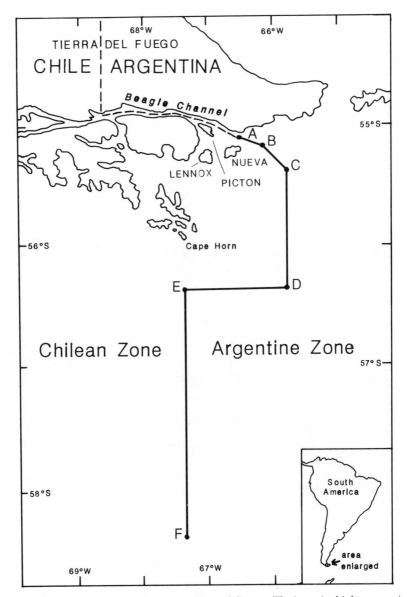

Figure 4.4 The settlement of the Beagle Channel dispute. The issues in this long-running dispute were unusual in the rest of the world but familiar in the Cono Sur (southern cone) of South America. Both Chile and Argentina claimed the islands of Picton, Nueva and Lennox in the eastern mouth of the Beagle Channel. The islands themselves are of little value to anyone, there is little likelihood of important resources being developed in the region and the channel is seldom used for international navigation. The disagreement was largely geopolitical and psychological. Whichever State owned the islands would be entitled to an extensive EEZ and continental shelf. Chile wanted such a zone in the Atlantic to become a "bioceanic State"; Argentina wanted Chile confined to the Pacific. The papal arbitral award of 1983 awarded the islands to Chile, but greatly reduced the maritime jurisdiction that would have gone along with them. (Compiled from American and Chilean maps)

any kind of combination or other compromise. Instead, both Articles 74 (on the EEZ) and 83 (on the continental shelf) contain the identical wording. The operative clause for the four paragraphs is Paragraph 1. It provides that delimitation in these cases "shall be effected by agreement on the basis of international law, as referred to in Article 38 of the Statute of the International Court of Justice, in order to achieve an equitable solution."

Although most continental shelf and EEZ boundaries between opposite and adjacent countries will probably be settled through bilateral negotiation, others will have to be drawn with the aid of third parties. Indeed, already the bulk of the cases being considered by the International Court of Justice (ICJ or World Court) and by arbitrators in the arena of public international law involve these types of boundaries, a new version of a traditional problem in political geography. The first of these was the North Sea continental shelf cases, settled by the ICJ in 1969. Since then others have involved Greece and Turkey, Libya and Tunisia, Libya and Malta, Guinea and Guinea Bissau, the UK and France, Argentina and Chile in the Beagle Channel, the USA and Canada in the Gulf of Maine and most recently Denmark and Norway in the area between Greenland and Jan Mayen Island.

Besides the questions of principles to be applied and resources to be allocated, maritime boundary disputes may arise out of technical problems left unresolved by the Convention. Parenthetically, we may note that map scales and projections are not specified, neither are sea level datums nor, as we pointed out earlier, just what is meant by "low-water mark." An example of the kind of politicogeographic problem that can arise because of these technical deficiencies in boundary provisions of a treaty is the dispute between the USA and the USSR over their boundary through the potentially oil- and gas-rich Navarin Basin just south of the Bering Strait. The two parties cannot agree on whether the line specified in the 1867 Convention transferring Alaska from Russia to the USA was a rhumb line or part of a great circle, and the original maps used in 1867 have been lost. Clearly, we are at the beginning of an era in which maritime boundary problems will be numerous, important and difficult.

5 Special zones and regimes

As comprehensive as it might seem, Chapter 3 did not cover several situations of great interest to political geographers, and for the most part of great importance to the world. These deserve particular attention here. Among them, the most important by far – at least at this stage of history – is the question of straits.

Straits used for international navigation

Traditionally, passage through and over straits around the world has been free, uncomplicated and unimpeded except in wartime. Community interests clearly demanded unfettered movement of civilian cargos and the major naval powers insisted on – and assured – free movement of their wide-ranging warships. The 1958 Geneva Conventions paid scant attention to the subject; it is covered, in fact, by only one article in the Territorial Sea Convention. This is no indication of its importance, however, only of its relative simplicity in an era in which few straits used for international navigation came completely under national jurisdiction. With the expansion of territorial seas from three, four or six miles to twelve, however, the high seas in scores of straits vanished and free passage through them became problematical.

The key rule in the Territorial Sea Convention bears verbatim examination:

> There shall be no suspension of the innocent passage of foreign ships through straits which are used for international navigation between one part of the high seas and another part of the high seas or the territorial sea of a foreign State.

Note that:
a) There is no definition of "strait" in the Convention, and therefore its ordinary meaning is accepted;
b) The phrase used is not "international strait," which would have excluded

straits entirely within one country, such as those between islands of Indonesia or of the Canadian Arctic archipelago;

c) Attempts to modify "used for international navigation" by adding "normally", "historically" or "traditionally" were rejected, thus making *all* straits subject to this rule, regardless of how seldom they may actually be used for this purpose, but only those portions comprised of territorial waters; in broader straits that had high seas, or international waters, through them, the high seas freedom of passage applied;

d) The applicable rule here is that of innocent passage, as in other territorial waters, but in straits it is non-suspendable, though normally a State may suspend innocent passage or limit it temporarily for security reasons; there is no innocent passage right for aircraft, even over straits; and

e) The last clause, "between one part of the high seas and another part of the high seas or the territorial sea of a foreign State," was known as "the Israel clause" because it applied specifically to the Strait of Tiran, blockaded by Egypt from 1949 and opened by force during Israel's Sinai campaign less than two years before UNCLOS I. At that time, the Strait of Tiran was being kept open for ships of all States and for ships bound to and from Israel by UN peacekeeping forces posted at Sharm el Sheikh; this clause was meant to reinforce the traditional freedom of navigation.

During the next decade, as more and more countries claimed broadening belts of national jurisdiction, the US and the USSR, the world's leading maritime powers, became more concerned about possible restrictions on their navies' freedom of movement, especially through "choke points," critical straits through which major shipping lanes funnel because no viable alternative routes exist. Their consultations on this question proved to be one of the forerunners of UNCLOS III. The two countries went into the new Conference determined to protect maximum freedom of navigation and throughout the early stages of the Conference worked together to achieve this goal. They were not entirely successful, in view of the opposition of some key straits States and others of the Group of 77. The US proposal for a "corridor of high seas" through all straits, for example, was rejected early on, and the ultimate compromise result was an entirely new concept in the Law of the Sea: transit passage.

Part III of the United Nations Convention on the Law of the Sea consists, not of one article on straits used for international navigation, as in 1958, but 12. Of the 12, eight spell out the new regime of transit passage. It is a real and not just a semantic compromise between the free transit demanded by the maritime powers and the innocent passage demanded by most of the smaller, poorer countries. It differs from free transit in that ships passing through straits may not do as they please, but are subject to a list of restrictions similar to those applicable to innocent passage through the territorial sea. It differs from the latter, however, in several important ways.

First, it does not require submarines to transit straits on the surface with

Figure 5.1   Vital straits used for international navigation. This satellite photograph of the southern tip of the Malay Peninsula shows the Straits of Malacca (lower left) and Singapore (lower center), two straits through which pass immense quantities of petroleum and other goods of all descriptions. Singapore is now the world's leading seaport, having passed Rotterdam several years ago. Between Singapore Island (center) and the mainland is Johore Strait, used primarily by local traffic. (NASA)

their national flags flying, as is required in territorial waters. Second, it applies to aircraft, which have no general right of passage over territorial waters. Third, Israel, if the regime of transit passage is accepted as valid international law, must depend upon faithful execution of relevant clauses in its peace treaty with Egypt to keep the Strait of Tiran open to its shipping, or upon a favorable interpretation of Article 45, which preserves the regime of innocent passage in certain circumstances, the only remnant of the 1958 "Israel clause" in the 1982 Convention. Finally, while the transit passage regime is fully as elaborate as that of innocent passage, there are a number of differences relating to the rights and duties of both vessels and the straits States, some of which are ambiguous and remain to be tested in practice.

In fact, the entire regime of transit passage is still being hotly debated in naval circles and in the legal literature. Whether it will, in fact, function as an equitable balancing of the rights and needs of straits States and international navigation, and even whether it will be accepted as valid customary law before the LOSC enters into force or if it never enters into force, remain to be seen.

The regime of islands

As children, we all learned that an island is "a piece of land surrounded by water," and the definition seems quite simple and perfectly adequate. In fact, however, the long debates at UNCLOS III and earlier Conferences over "What is an island?" would have been both amusing and dismaying to the average layman. But the question is neither amusing nor academic, especially now, when a flyspeck in the vast ocean, such as Rockall in the northeast Atlantic with a circumference of only about 100 meters (330 feet) could be endowed with an economic zone of 125,000 square nautical miles! Then consider that there are more than 500,000 oceanic islands in the world, ranging downward in size from Greenland, bigger than all but 11 countries, to thousands the size of Rockall and even smaller, many visible only at low tide or otherwise of doubtful status. Clearly, then, a legal definition is necessary.

The 1982 Convention, as might be expected, elaborates upon the simple 1958 Territorial Sea Convention definition. Part VIII of the new Convention consists of one article (121), given here in its entirety:

1 An island is a naturally formed area of land, surrounded by water, which is above water at high tide.
2 Except as provided for in paragraph 3, the territorial sea, the contiguous zone, the exclusive economic zone and the continental shelf of an island are determined in accordance with the provisions of this Convention applicable to other land territory.
3 Rocks which cannot sustain human habitation or economic life of their own shall have no exclusive economic zone or continental shelf.

Paragraph 1 reproduces the 1958 definition; the other two paragraphs are new. All three, however, present difficulties. The definition, for example, does not take into account ice islands, permafrost islands or spoil piles made of "natural" material deposited by man but "formed" by the action of wind and sea. And if a country constructs some breakwater or other structure in the sea in such a fashion that wind and tide pile up sand around it, is the resultant island a "naturally formed" or an artificial island? These questions are important, for elsewhere the Convention (Arts 60(8) and 80) says clearly, "Artificial islands, installations and structures do not possess the status of islands. They have no territorial sea of their own, and their presence does not

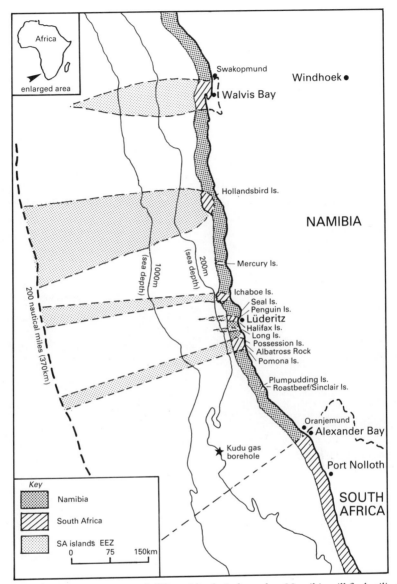

Figure 5.2 The importance of offshore islands. Independent Namibia will find utilization of her maritime space hampered not only by the continued possession of the fine port of Walvis Bay by the Republic of South Africa, but also by South Africa's screen of islands close to the Namibian coast. Note how these islands break up Namibia's territorial sea (the 12-mile band along the coast), but also, in four significant places, her exclusive economic zone. This situation is similar to that of the Greek Aegean islands near Turkey's coast and may also lead to a prolonged dispute. Another potential problem is the maritime boundary between Namibia and South Africa, shown here in a highly generalized way, which remains to be negotiated.

affect the delimitation of the territorial sea, the exclusive economic zone or the continental shelf."

Paragraph 2 is the critical part of this article, for it means that now islands (however defined) formerly so small, remote and useless that they were visited only rarely by the occasional fisherman, and generally avoided as hazards to navigation, have suddenly acquired enormous potential value because of the riches that may lie in or beneath the sea around them. This has already led to many sovereignty disputes in nearly all the oceans of the world and many more are likely to arise.[1]

In an effort to reduce disputes over "What is an island?", the negotiators at UNCLOS III adopted paragraph 3. But this formulation itself is a compromise and hardly unambiguous. The terms "rock," "human habitation" and "economic life" are not defined. Can a lighthouse keeper on a lonely rock, for example, be considered "human habitation" even if he grows his own food in a tiny garden or with hydroponics? Is a weather station or a pollution monitoring station "economic life"? Not only does paragraph 3 fail to answer the question of which "islands" are entitled to EEZs and continental shelves, but also of which are to be considered in drawing baselines or incorporated into territorial seas. It is no longer possible to ignore *any* geographical or geological feature on the surface of the earth, no matter how small and insignificant it may seem, unless all interested parties agree to do so.

Coral reefs of atolls present more problems, particularly now that so many tropical and subtropical island territories have become sovereign States. While discussed to some extent by the International Law Commission and in the professional literature earlier in the post-World War II era, they were not mentioned in the 1958 Conventions or elsewhere in the Law of the Sea. The 1982 Convention does address them in Article 6, providing that the seaward low-water line of a fringing reef may be used as the baseline from which the breadth of the territorial sea is measured. There are still ambiguities in this provision, and some latitude for States in applying it, but so far it does not seem to have created any serious problems.

The regime of islands is one more example of the increasing complexity of formerly simple matters in the political geography of the sea. Like so many other topics we consider in this book, it is likely to keep lawyers and geographers busy for a very long time. Yet it is still relatively simple compared with archipelagoes.

Archipelagic States

Again, we learned in childhood that an archipelago is a group of islands, and that definition is still satisfactory for physical geographers. But political geographers must grapple with the problems generated by such an elemental definition as a result of the new Law of the Sea. For more than half a century, since before the Hague Conference of 1930, the status of archipelagoes has been a contentious issue. At first the question was basically whether offshore islands, or "coastal archipelagoes," could be enclosed by straight baselines from which the breadth of the territorial sea could then be measured. Such countries as Norway, Denmark and Yugoslavia were most active in the debate through 1958. The question was partially answered by the World Court in the Anglo-Norwegian Fisheries Case of 1951 and its ruling was in its essence incorporated into the Territorial Sea Convention (as discussed in Ch. 4), but this was far from the end of the controversy.

At the 1958 Conference, Indonesia and the Philippines led the fight for a special regime for oceanic rather than coastal archipelagoes. They had already (in 1957 and 1955 respectively) announced that they would consider as internal waters all of the sea enclosed by straight lines connecting the outermost points of their outermost islands. Their task, then, was not only to justify and legitimize this action, but to have the principle expanded into a regime recognized by the international community as essential for the maintenance of the integrity of their territories. They argued – with considerable justification – that the waters among their islands were historically, economically and culturally as much a part of their national territory as the islands themselves, and that in order to preserve this unity, they needed total political and legal control over them.

Although there was much sympathy for this position, the maritime powers feared a loss of navigational rights in these waters were such a regime to be created. In the end, the maritime powers won, and mid-ocean archipelagoes were not granted any dispensation from the rules adopted for drawing straight baselines. Most of the waters within the archipelagoes would remain, then, either territorial seas or high seas, with few restrictions on navigation.

After 1958, however, more archipelagic territories became politically independent and increasingly dependent on fishing for their economic health. Pollution of the sea became a serious problem; so did smuggling, illegal immigration, security and vastly increased inter-island traffic. By the time of Caracas in 1974, Indonesia and the Philippines had been joined by Fiji and Mauritius in the fight for an archipelagic regime. By working very hard, by marshalling convincing arguments, and through negotiating skill, they were finally able to prevail – but not entirely. The maritime States were also determined and skillful, and managed to preserve much of their highly valued freedom of navigation. The result was a regime of great complexity and many uncertainties, much like that of straits. Indeed, the two might be considered together, with many of the relevant provisions of the Convention lined up in parallel columns to illustrate the similarities. There are important differences, however.

First, as always, is the matter of definitions. Part IV of the Convention is headed "Archipelagic States", and consists of nine articles, two of them very long. Article 46 lays out definitions:

For the purposes of this Convention:

(a) "archipelagic State" means a State constituted wholly by one or more archipelagoes and may include other islands;

(b) "archipelago" means a group of islands, including parts of islands, interconnecting waters and other natural features which are so closely interrelated that such islands, waters and other natural features form an intrinsic geographical, economic and political entity, or which historically have been regarded as such.

These definitions clearly exclude such countries as Canada, Equatorial Guinea, Greece and the USA which consist not only of archipelagoes but also of mainland territories. Many States consisting wholly of islands are also excluded from the category of "archipelagic State" by the complex provisions of Article 47, which spell out in great (but probably not sufficient) detail the methods of drawing archipelagic baselines. These provisions include water-to-land ratios, lengths of baseline segments, rules on low-tide elevations, a requirement to follow "the general configuration of the archipelago," etc. Such countries as Japan, Kiribati, Tuvalu and the UK are among those excluded. The Bahamas, Fiji, Indonesia, the Philippines, and a very few others clearly qualify as archipelagic States. Other island States may or may not qualify depending on the interpretation and application of Article 47.

Once the principle of a special regime for archipelagic States had been accepted by the maritime powers, their major efforts were concentrated on, first, limiting as much as possible the number of States that could qualify and, second, assuring themselves the greatest possible freedom of navigation in archipelagic waters. The results of the first effort are evident in Article 47, one of the very long articles in this part, and of the second in the other very long one, Article 53.

Although other articles provide for the application of the rules already discussed in Chapter 3 pertaining to internal waters and innocent passage, and for a new maritime zone, archipelagic waters (with the airspace over them and their seabed and subsoil), Article 53 is crucial to the entire compromise. It provides that: "An archipelagic State may designate sea lanes and air routes thereabove, suitable for the continuous and expeditious passage of foreign ships and aircraft through or over its archipelagic waters and the adjacent territorial sea." Archipelagic sea-lanes passage, as described in Article 53, is very similar to transit passage through straits and thus includes navigational rights more extensive than those provided under the innocent passage rules applicable in the territorial sea, which lies *outside* the baselines delimiting the archipelagic waters.

It has been argued that archipelagic waters are unnecessary, since an island State could achieve essentially the same results simply by declaring a territorial sea and where possible an EEZ around each individual island. But this very provision for sea lanes passage through and over archipelagic waters has no counterpart in the EEZ, where high seas freedoms of navigation prevail. As a practical matter, this is the price the archipelagic States had to pay for the psychological value of universal recognition of the unity of their islands and the sea among them.

If all of the provisions of Part IV are faithfully carried out by affected countries, the new regime of archipelagic States with its right of archipelagic sea-lanes passage could bring tranquility to some hitherto or potentially troubled waters. There is no reason to believe, however, that this will be the case. Already, a number of States, including the Cape Verde Islands, Denmark, Indonesia, the Philippines and Spain, have apparently violated one or another of the provisions, and more may be expected to do so. Whether any of them will eventually conform if and when the Convention enters into force (or before then) or whether such violations really matter in practical terms can only become evident in time. Meanwhile, the archipelagic State is an interesting new addition to the existing categories of States and will also provide topics for theses and dissertations – and work for geographers and lawyers – for many years to come.

Enclosed or semi-enclosed seas

One of the most peculiar, even mysterious, elements of the United Nations Convention on the Law of the Sea is Part IX, "Enclosed or Semi-enclosed Seas". The mystery is why it appears in the Convention at all. It consists of only two articles, neither of which has any discernible value.

Article 122 defines the term as meaning "a gulf, basin or sea surrounded by two or more States and connected to another sea or the ocean by a narrow outlet or consisting entirely or primarily of the territorial seas and exclusive economic zones of two or more coastal States." Equating "enclosed" and "semi-enclosed" seas makes no sense. If a sea has an outlet it is not enclosed. The only enclosed sea in the world generally regarded as such is the Caspian and it was long ago partitioned between Iran and the USSR and is not covered by the Law of the Sea. All the other gulfs, basins or seas may be semi-enclosed or largely open. While semi-enclosed seas, such as the Caribbean, Mediterranean and South China seas, certainly have some common characteristics and some that set them apart from the open ocean, none of these characteristics appears to create circumstances warranting any special treatment in the Convention. All conceivable circumstances in such seas are adequately covered in other parts of the Convention.

Article 123 merely says that, "States bordering an enclosed or semi-enclosed sea should co-operate with each other in the exercise of their rights and in the performance of their duties under this Convention." It then goes on to list four areas of such suggested cooperation. But *all* States should cooperate in these and all other areas, and suggestions, exhortations and commands to cooperate are found prolifically throughout the Treaty. In fact, such cooperation so far seems to be more evident among States *not* facing semi-enclosed seas, with the notable exception of those participating in the Oceans and Coastal Areas Programme of the United Nations Environment Programme (UNEP), which developed outside the framework of the Law of the Sea.

There is little in either the UNCLOS III documentation or the professional literature to explain Part IX. Apparently, the USA, USSR and other maritime States saw to it that no special regime was established for semi-enclosed seas that would be likely to hamper their mobility in these constricted areas. Concern about such a possibility seems to have been generated by a "think piece" by a prominent US political geographer, Lewis Alexander.[2] He did not propose any particular regime for semi-enclosed seas, but he did offer a lucid discussion of them in the context of the need for marine regions in a new era of the Law of the Sea.

In 1988 one of the proposals for resolving some fisheries problems in the Bering Sea was to consider it a semi-enclosed sea and have the US and USSR jointly administer the "doughnut hole," the small area of high seas between their EEZs in the Bering Sea. There was not much support for this proposal, however, and as a practical matter Part IX remains dormant. It is possible that some day Part IX will assume an importance we cannot envisage now, but for the present it is of little consequence.

6 Land-locked States

There are at present some 30 land-locked States in the world (depending on one's definition of "State"), half of which are listed by the United Nations as "least developed," the poorest of the poor, while all but the nine in Europe are poor in varying degrees. This poverty is no accident, for a country without a seacoast is severely handicapped.

The lack of a seacoast is a handicap for several reasons. First, sheer distance from a seaport increases the costs of transporting both imports and exports of the land-locked State. These costs are multiplied by such factors as difficult terrain, adverse climates and inadequate transportation routes and equipment. Since most of the countries across which developing land-locked countries must transit are themselves poor, they frequently find their ports and transport facilities inadequate for their own expanding needs, to say nothing of having a surplus capacity for the use of a land-locked neighbor. Delays in delivering transit traffic of land-locked States often result from cumbersome procedures, shortages of competent staff, inadequate and unreliable communications, and similar problems. These delays increase costs through loss, damage, deterioration and pilferage; higher insurance rates, credit charges and demurrage; and sometimes higher prices charged by exporters to inland countries.

Many of these problems are also encountered by the interior districts of coastal States, but they do not have to deal with the additional difficulties of crossing international boundaries. Land-locked States, moreover, suffer disruptions of transit traffic due to strikes, civil disturbances and war in the transit States, even though they themselves are not directly involved.

As the developing land-locked countries attempt to expand their economies, international trade becomes more important to them and these impediments to trade become more costly and troublesome. Finally, seaports have always served as "windows on the world" through which flow people and ideas, as well as goods, to stimulate change and growth; countries which lack seaports are therefore inhibited in their efforts to modernize.

From ancient times, land-locked territories have often faced obstructions, restrictions, tolls or heavy transit fees on goods and persons en route to or from the sea. Gradually, however, insistence on absolute sovereignty by

Figure 6.1 The weak link in Laos' access to and from the sea: a ferry across the Mekong River. The plight of many developing land-locked countries, especially the poorest of them, is illustrated by this primitive ferry which is the chief link between Laos (on the far shore) and Thailand, Laos's most important transit State. An old proposal to replace this ferry by a rail-cum-highway bridge between Nonghai in Thailand and Vientiane in Laos has recently been revived, but is unlikely to be realized very soon. Meanwhile, Laos is dependent on transport such as this for her international trade. (Martin Glassner)

transit States began to give way to a recognition of the advantages of a free flow of trade. During the 19th century principles of free transit became established in Europe as recognition grew that such transit was necessary for the development of commerce and industry to the benefit of both land-locked and transit States.

The League of Nations sponsored a series of conferences during the 1920s which produced both bilateral and multilateral treaties aimed at the facilitation of free transit. Of these, the most important were the Barcelona Convention and Statute of 1921 and the Convention and Statute of the International Regime of Maritime Ports, signed at Geneva in 1923, which set minimum standards for the transit and other rights of the land-locked.

After World War II, several important events and trends conjoined to improve access to the sea for land-locked States. For example, territorial reorganizations in Europe, which eliminated the Polish and Finnish Corridors, reduced pressure for the creation of new corridors to provide sovereign outlets to the sea and enhanced international cooperation in both overland and water transportation. An elaborate system of internationalized rivers and canals, free zones and free ports, and special transit arrangements has made Europe today important in discussions of access to the sea largely as an example of how land-locked and transit States can develop harmonious and mutually satisfactory arrangements.

During the last century, a relatively new concept has developed, that access to the sea is essential for the expansion of international trade and economic development. The achievement of independence by many former colonies in Asia and Africa stimulated increasing emphasis on this relationship, rather than the traditional stress on "rights and duties." Afghanistan, for example, whose access to the sea through Pakistan had been cut off in 1950, 1955, and again in 1961–3, was a leader in the battle for a "free and secure access to the sea." So was Bolivia, which was trying to win back some or all of the seacoast province which it lost to Chile in the War of the Pacific (1877–83). These efforts produced Article 3 of the Convention on the High Seas in 1958. This represented progress, but pressure increased for a more definitive answer to the question.

As a result, the 1964 United Nations Conference on Trade and Development (UNCTAD) at Geneva adopted eight principles on access to the sea for land-locked States, and several recommendations, including one for a full-dress conference on the subject. The conference was held in July 1965 and was successful.

The 1965 United Nations Convention on Transit Trade of Land-locked States, while certainly giving a measure of "status" to the problems of access to the sea for developing land-locked States, did not solve them. Its effectiveness is limited, moreover, by the fact that only a few transit States have ratified or acceded to it and only one of these (Nigeria) provides transit for *developing* land-locked States, in this case Niger and Chad. Despite its importance as a yardstick, it was simply another step in the long drive of developing land-locked States to reduce to a minimum the difficulties inherent in their mediterranean ("in the middle of the land") location.

Since 1965 they have continued to strive in this direction in a variety of ways which can be grouped into five major categories:

1   Internal development, particularly of transportation systems;
2   Bilateral negotiations for improved transit;
3   Practical measures to improve transport facilities in transit States and also develop alternative routes to the sea;
4   Regional economic integration; and
5   Continuing efforts in the United Nations and its specialized organs to establish and implement a right of transit.

All of these activities deserve attention, but our limited space compels us to concentrate only on a sixth approach used by land-locked countries to obtain and protect their access to the sea: the development of international law, particularly in UNCLOS III. Here, because of the many changes that had taken place in the world since World War II, particularly decolonization and the extension of national jurisdiction, the land-locked countries had to fight another battle – for access to the *resources* of the sea. In these efforts, they made an alliance with a number of countries that, for one reason or another, considered themselves "geographically disadvantaged" because they were able to gain little or nothing from the extension seaward of national jurisdiction.

The land-locked States at UNCLOS III

As we indicated in Chapter 2, one of the largest, most diverse, and most pesistent interest groups at the Third UN Law of the Sea Conference was the Group of Land-locked and Geographically Disadvantaged States. Its roots lay in the Group of Land-locked States, formed at the initiative of Afghanistan at the Eleventh Session of the United Nations General Assembly in 1957. It held its first formal meeting in Geneva, 10–14 February 1958, at the invitation of Switzerland, to prepare for the first United Nations Conference on the Law of the Sea in April. That group produced documents on transit that were considered by the Fifth Committee of the 1958 Conference, but the end result, Article 3 of the Convention on the High Seas, was not commensurate with either the needs of land-locked countries or their effort. Article 3 merely provided that land-locked States and their States of transit should settle all matters relating to freedom of transit "by mutual agreement."

The land-locked States also failed in their efforts to have written into official drafts of the UN Seabed Committee after 1969 their proposal for "rights" of free and unrestricted access to the sea, failed to gain participation in the EEZs of coastal States "on an equal and nondiscriminatory basis," and failed to obtain provisions for regional economic zones. Opposed to Venezuela's 1971 proposal for coastal State "patrimonial seas," in which the coastal States would be virtually sovereign, they were, nevertheless, maneuvered into accepting the concept of an exclusive economic zone without getting anything in return. Outside the Seabed Committee, meanwhile, they were also being outmaneuvered in a series of regional meetings in Latin America and Africa before the Caracas session of UNCLOS III in 1974. The African EEZ proposal, though, would have allowed land-locked States to share in the *living* resources of such zones off their own continents.

At Caracas the Land-locked Group expanded by admitting both *developed* land-locked States and "geographically disadvantaged" States. While adding to their negotiating and potential voting strength, these moves created considerable difficulties in formulating positions for what had been exclusively a group of developing land-locked States. The difficulties were to grow and be reflected in changing positions taken by the group and in its inability after 1976 to take a firm position on any specific issue of consequence. Nevertheless, bizarre as it was, the Group ultimately reached a peak strength of 55 members in March 1981 and survived until the end of the Conference in December 1982. The Convention that UNCLOS III adopted contained features both helpful and harmful to the land-locked States.

Convention provisions on land-locked States

Table 6.1 provides a framework for analysis of the 1982 Convention articles relating to land-locked States and serves as a ready reference for locating various types of provisions in the Convention.

Of the six articles in the two sections of Part XI that relate to land-locked States, only Article 161 has elicited serious complaints from them. It provides that "States which are land-locked or geographically disadvantaged" are one of the categories of "developing States, with special interests" which may appoint six of the 36 members of the International Seabed Authority's Council. Some land-locked States fear that in practice they will be under-represented in the Council, but this has not been a serious concern and is unlikely to be for many years.

All of the remaining articles, except for Articles 69 and 254 and Part X, were not particularly controversial at the Conference and are generally satisfactory to land-locked States. Their deficiencies, generally in the form of vagueness, are not serious enough in a convention not yet in force to warrant intensive discussion.

The land-locked States fought against the extension of coastal State jurisdiction farther out to sea. They lost access to the continental shelf early on, however, because of the vigor with which coastal States espoused the continental shelf doctrine. The only major question was a legal definition of the outer limit of the shelf. The land-locked States fought and lost the battle for a narrow shelf definition, but they did help to win "payments or contributions in kind in respect of the exploitation of the non-living resources of the continental shelf beyond 200 nautical miles . . . " under Article 82. But it may be at least a quarter century before any land-locked State actually receives any significant compensation for the loss of access to the great mineral resources of this area.

The coastal States were from the beginning absolutely adamant in refusing to consider any sharing of revenues from, or any access to, the "non-living" resources of their exclusive economic zones. The Group of Land-locked and Geographically Disadvantaged States was simply unable to get a portion of the resources of the area that contains all of the world's known offshore oil and gas, all the presently exploitable minerals, the most readily accessible polymet-allic nodules, and most of the known deposits of polymetallic sulfides. All the coastal States would grudgingly concede to the land-locked States was a

Table 6.1 1982 Convention articles relating to land-locked States

	Non-discrimination	Special needs and interests	Specific provisions
Preamble		Preamble	
Part II Territorial Sea and Contiguous Zone	17		
Part V Exclusive Economic Zone	58		69
Part VI Continental shelf		82	
Part VII High seas	87,90		
Part X Access to the sea/transit			124–132
Part XI International Seabed Area	140,141	148	
Part XI International Seabed Authority		152,160	161
Part XIII Marine scientific research	238		254
Part XIV Transfer of marine technology	274	266,269,272	
Annex IV Statute of the Enterprise	12		

greatly circumscribed access to a very small portion of the "living resources" of the exclusive economic zone.

Article 69, Paragraph 1 by itself offers little enough to the land-locked countries. They are only granted "the right to participate, *on an equitable basis*, in the exploitation of an *appropriate* part of the *surplus* of the living resources of the exclusive economic zones of coastal States *of the same subregion or region*, taking into account *the relevant economic and geographical circumstances* of *all* the States concerned...." (emphasis supplied).

But Article 69, Paragraph 1 does not stand alone. Paragraph 2 lists four factors that must be taken into account in establishing the terms and modalities of the participation of land-locked States in resource exploitation through bilateral, subregional, or regional agreements. The coastal States are clearly given the upper hand. Paragraph 3 requires that agreements be reached permitting the participation of land-locked States in exploitation of a prospective surplus "as may be appropriate in the circumstances and on terms satisfactory to all parties," taking into account the four factors mentioned in Paragraph 2.

Finally, a *developed* land-locked State can only participate in exploiting the exclusive economic zone of a *developed* coastal State, with additional restrictions. In general, although the grant of "equal or preferential rights for the exploitation of the living resources in the exclusive economic zones" is permitted, it is not encouraged.

These restrictions alone assure very little benefit from the exclusive economic zone for any land-locked country, but to them must be added other considerable limitations contained in other articles of the Convention. The land-locked States were simply thrown a few fish to help avoid their rejecting the 1982 Convention as a whole.

Article 254, which deals with the rights of land-locked and geographically disadvantaged States in regard to marine scientific research, requires that they be notified by States proposing research projects within the jurisdiction of neighbouring coastal States; that after consent by the coastal State for such projects is given, they be provided relevant information "at their request and when appropriate," and that they may participate, "whenever feasible," in the proposed marine scientific research project through qualified experts appointed by them "*and not objected to by the coastal State....*" (emphasis added).

Some coastal countries objected to land-locked countries being given the *right* to participate in any way in marine scientific research in any areas under coastal State jurisdiction. The land-locked and geographically disadvantaged States, however, had strongly supported a liberal regime for marine scientific research and this article was, in a sense, their reward for this support.

Marine scientific research, moreover, benefits the entire international community and no State should be prohibited from engaging in it. There is ample protection in the 1982 Convention for the economic and military security of the coastal State, so participation in such research by land-locked States under the same terms and conditions that apply to all other participants should pose no threat to any coastal State. Besides, land-locked States could, on their own or through "qualified experts appointed by them," make significant contributions to our understanding of the sea.

Part X of the Convention deals with the right of access of land-locked States to and from the sea, and freedom of transit. This part essentially reveals the story of the losing struggle of the land-locked States at UNCLOS III, as in previous conferences, to firmly establish in international law a *right* of transit. The coastal States adamantly refused to change the word "freedom" to "right," arguing that transit across their territory was not a right granted by natural law or some other higher power, but a privilege graciously granted by them individually, which may be modified or withdrawn at any time; otherwise it would be a violation of their sovereignty.

The nine articles in Part X can best be analyzed by comparing the most important of them with both the 1965 United Nations Convention on Transit

Figure 6.2 The port of Matarani, Peru. Matarani was built in the 1960s to replace obsolete Mollendo as the chief port for southern Peru. A large part of its traffic, however, consists of tin and other ores shipped from Bolivian mines across Lake Titicaca on modern railcar ships and then railed down to the port. Matarani and Antofagasta, Chile, are Bolivia's principal outlets for her exports, while imports pass primarily through Antofagasta and Arica, Chile. (Martin Glassner)

Trade of Land-locked States (negotiated and signed at UN Headquarters in New York), and what might be considered ideal convention articles for the benefit of land-locked States. Some of the suggestions for improvement of the Convention articles offered below were introduced into the negotiations in the Conference but rejected by the coastal States. Nevertheless, some have already been incorporated into bilateral transit treaties and others may be adopted in the future.

Article 125 concerns the Right of Access To and From the Sea and Freedom of Transit. This title is a distinct improvement over the phrasing of Article 3 of the 1958 Convention on the High Seas and of Principle III of UNCTAD I, incorporated verbatim into the preamble of the 1965 New York Convention, which says only that land-locked States "should have free access to the sea." In other respects, however, it is much weaker than Article 2 of the 1965 Convention, which requires that the transit State "shall facilitate traffic in transit. . . .".

Article 130, Measures to Avoid or Eliminate Delays or Other Difficulties of a Technical Nature in Traffic in Transit, is almost identical in wording to Article 7 of the New York Convention. There are, however, some differences in paragraph 1. The words "or eliminate" have been added to the title, but this phrase does not appear in Paragraph 1. Paragraph 2, however, requires cooperation toward the "expeditious elimination" of the delays or difficulties mentioned in Paragraph 1. This is made more difficult for the land-locked States by the insertion of the word "appropriate" to modify "all measures," a phrase that appeared unmodified in the 1965 Convention.

By far the most significant change in the 1965 formulation is the insertion of the phrase "of a technical nature" to modify the "delays or other difficulties" that the transit State must "avoid" and must cooperate in expeditious elimination. If interpreted rigidly, a transit State could argue that it has no obligation at all to avoid delays, because none is merely "technical." But an argument that would ignore the other words of the article is inherently unconvincing.

All told, nine articles of the Convention on Transit Trade of Land-locked States have no equivalent provisions in Part X of the new Convention. Of them, three are simply unnecessary now, three would be good for the land-locked countries and three would be bad. The good ones spell out a number of very important obligations of the transit State, all designed to permit "the carrying out of free, uninterrupted and continuous traffic in transit" in the words of Article 5, Paragraph 1. Certainly they should be included in some form in any ideal transit treaty and Part X would be much more useful to the transit States, as well as to the orderly and efficient flow of world commerce in general if they were included.

Of the three articles dropped between 1965 and 1982 which were harmful to the interests of land-locked States, two allowed the transit State to deviate from the provisions of the Convention and restrict transit under a variety of circumstances. The third, Article 15, reads, "The provisions of the Convention shall be applied on a basis of reciprocity." The battle in the UN Law of the Sea Conference over this provision was intense, and it resulted in one of the more important of the modest victories of the land-locked States in the Conference. It is debatable, however, whether this gain will offset the loss of the commitment of the transit State to the "free, uninterrupted and continuous" flow of traffic in transit.

Comparing the position of land-locked States overall in regard to the Law of the Sea before and since UNCLOS III, it is safe to summarize by saying that in terms of transit to and from the sea their situation is now slightly better; in respect of uses of the sea, however, they have suffered a serious loss. The gravity of the loss will only become evident as their economic development both demands and permits greater use of the sea. By that time, however, it is possible that regional arrangements will have been worked out that will allow the land-locked countries of the world – developed and developing – access to the resources of the sea on fair and reasonable terms. If so, such arrangements will help significantly to reduce the possibility of conflict over the riches of Neptune's domain.

7 Protecting and preserving Neptune's domain

One of the weaknesses of the United Nations Convention on the Law of the Sea, especially from a geographer's perspective, is its failure to treat the sea as "an indivisible ecological whole."[1] Instead, the world ocean, the global sea, has been partitioned – horizontally, vertically and functionally – into innumerable international, regional, national and private jurisdictions with no centralized control over them, and no overall plan for preserving its integrity. Only once in the Convention is the word "ecosystem" used (in Art. 194(5)), referring to measures "necessary to protect and preserve rare or fragile ecosystems as well as the habitat of depleted, threatened or endangered species and other forms of marine life"). And there is no reference at all to ecology, the interaction of living things with their natural environments.

This unfortunate situation is undoubtedly due to the highly political organization of the Seabed Committee and UNCLOS III and to the emphasis by delegates on the division and exploitation of resources and secondarily on preserving freedom of navigation. These two objectives are often incompatible and neither, given the scale and intensity of such activities today, is compatible with the preservation of the marine environment as a whole. As a result, at UNCLOS III aspects of marine ecology were considered in all three main committees and as a subhead under a variety of subjects. The low priority assigned to the matter originally was at least in part due to the slight concern of the developing countries with environmental matters. Initially, they considered it a problem solely of the rich, industrialized countries which, after all, were the ones causing the greatest damage to the environment.

As the Conference progressed, however, they began to appreciate that they, too, were already causing environmental damage, and that their capacity to do so will increase as their populations increase and industrialization proceeds. In addition, since our entire planet constitutes one single ecosystem, environmental damage, no matter where it originates, affects everyone and all parts of the earth. This lesson was driven home by a series of marine oil spills around the world during the Conference, notably the discharge of some 210,000 tons of crude oil into the sea off Brittany by the damaged and drifting Liberian tanker *Amoco Cadiz*. This occurred in March of 1978, while UNCLOS III was in session in Geneva, not very far from Brittany. That day most of the delegates in the Third Committee conference room were reading newspaper accounts of the disaster, and very quickly the position of the French government almost reversed from indifference, even hostility, toward environmental issues, to support for much stronger treaty provisions.

This process of education and change of position during the deliberations of the Seabed Committee and UNCLOS III was a continuation of the world's growing concern with the health of the marine environment. This first became manifest at an international conference of experts on pollution of the sea by oil from ships, held at Washington in 1926 at the request of the United States. It produced a draft treaty, but this was never opened for signature. Interest in the subject waned after that. It was not covered at the Hague Conference in 1930 and Articles 24 and 25 of the 1958 Geneva Convention on the High Seas impose only general obligations on States to prevent pollution by oil and radioactive waste. Gradually, however, the International Maritime Organization (IMO, formerly the Intergovernmental Maritime Consultative Organization (IMCO) and a specialized agency of the United Nations with headquarters in London) developed a series of treaties, or conventions, beginning in 1954, that cover various aspects of the problem. States have been reluctant to accept and enforce restrictions designed to prevent pollution by ships, however, because they generally raise shipping costs.

But public awareness of actual and potential pollution of the sea was generated by such dramatic events as the grounding and sinking of the tanker *Torrey Canyon* off Land's End, England in 1967, the voyage of the US tanker *Manhattan* through Canada's Arctic waters in 1969, the leakage of oil from wells in the Santa Barbara Channel off the coast of California in 1969 and the voyages of Thor Heyerdahl across the Atlantic and Pacific on rafts from which he reported collecting lumps of crude oil floating far from any coast or shipping lane. Ordinary citizens in several countries worked through non-governmental organizations to pressure governments into action to prevent such pollution. Even shipowners and oil companies were finally galvanized into action to prevent more pollution disasters at sea. As a result, the relevant provisions of the LOSC are much stronger and more comprehensive than anyone originally expected them to be, but are still incomplete and weaker than many had hoped they would be.

Pollution of the sea

Part XII of the LOSC is titled, rather grandly, Protection and Preservation of the Marine Environment, and it begins (Art. 192) with "States have the obligation to protect and preserve the marine environment." The very next article, however, says, "States have the sovereign right to exploit their natural resources pursuant to their environmental policies and in accordance with their duty to protect and preserve the marine environment." The remaining 43 articles in this part (some of them very long) and the 14 other articles on the subject scattered through the text and its annexes continue this careful (even delicate) balancing of rights and obligations, interests and concerns. It is for this reason that the subject is of interest to political geographers and not just to environmentalists and physical geographers.

" Pollution of the marine environment" is defined in Article 1 of the Convention as:

> the introduction by man, directly or indirectly, of substances or energy into the marine environment, including estuaries, which results or is likely to result in such deleterious effects as harm to living resources and marine life, hazards to human health, hindrance to marine activities, including fishing and other legitimate uses of the sea, impairment of quality for use of sea water and reduction of amenities.

This definition is a very slightly expanded version of the one developed by the Intergovernmental Oceanographic Commission, an arm of UNESCO, and the UN Group of Experts on the Scientific Aspects of Marine Pollution. It does not satisfy the demands of those who would outlaw the introduction by man of *any* substances into the sea because not enough is known about the sea to be able to predict the long-term damage that might be caused even by seemingly harmless substances. However, it is the best we can obtain right now, is a vast improvement over previous definitions, and is adequate to deal with contemporary and foreseeable future pollution problems.

Overwhelmingly, the Convention provisions concerning pollution, whether in Part XII or elsewhere, deal with vessel-source pollution. This is both the most visible and dramatic form of pollution and the least important, since it accounts for only some 10 percent of all the pollutants that enter the sea. Nevertheless, it is the only form of marine pollution specifically included in the Conference's mandate. The fact that any other sources of pollution are even mentioned is testimony to the seriousness with which most delegations viewed the problem during the latter stages of the Conference. The LOSC thus includes some provisions long advocated by non-governmental organizations and long resisted by governments, and some of the provisions reflect real understanding of the ecology of the sea.

Perhaps the most comprehensive environmental provision in the entire Convention is Article 145, Protection of the Marine Environment:

> Necessary measures shall be taken in accordance with this Convention with respect to activities in the [International Seabed] Area to ensure effective protection for the marine environment from harmful effects which may arise from such activities. To this end the Authority shall adopt appropriate rules, regulations and procedures for, *inter alia*:

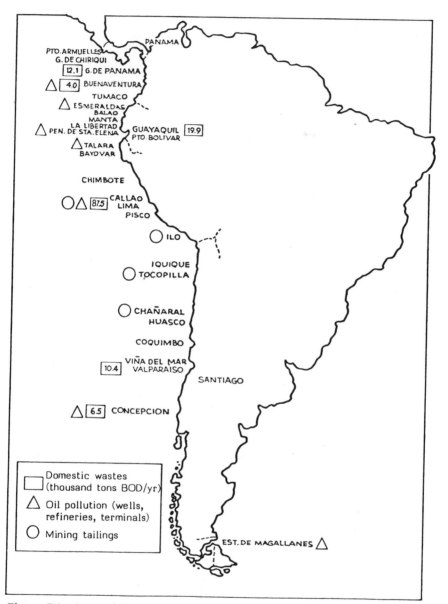

Figure 7.1 Areas of the southeast Pacific with major pollution problems. At the beginning of UNCLOS III the developing countries were little concerned with environmental problems in general and even less with pollution of the sea. By the end, however, they had come to understand that not only does pollution affect everyone, but also that everyone contributes to it. This region, which began industrializing fairly recently, already has serious marine pollution problems that are beginning to affect the fishing industry, so important here. That industry, too, contributes substantial waste from its fish canneries and fishmeal plants. (United Nations Environmental Programme 1985, p. 262)

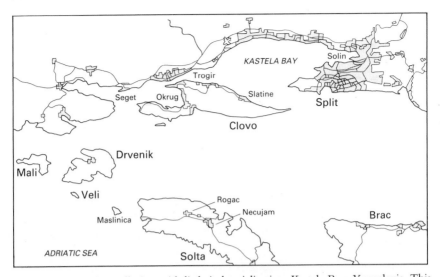

Figure 7.2 Marine pollution with little industrialization: Kastela Bay, Yugoslavia. This region on the central Dalmatian coast has little industry, no mining and only one modest-sized city (Split, population 240,000), yet it is afflicted by a serious marine pollution problem. This nearly enclosed bay has little circulation of water to mix and remove wastes. The major factor intensifying the normal municipal and industrial pollution of the bay is tourism. The population of 350,000 swells to some 565,000 in the summer. If the pollution gets too severe, the tourists will go elsewhere and the economy will suffer a heavy blow. The Yugoslav government has therefore proposed Kastela Bay as a pilot area for implementation of the Priority Action Programme of UNEP's Mediterranean Action Plan. (UNEP 1988)

(a) The prevention, reduction and control of pollution and other hazards to the marine environment, including the coastline, and of interference with the ecological balance of the marine environment, particular attention being paid to the need for protection from harmful effects of such activities as drilling, dredging, excavation, disposal of waste, construction and operation or maintenance of installations, pipelines and other devices related to such activities;

(b) the protection and conservation of the natural resources of the Area and the prevention of damage to the flora and fauna of the marine environment.

The wide scope of this article may be due to two factors. First, it applies to the deep seabed beyond the limits of national jurisdiction, a region that is essentially a *tabula rasa*, in which no State has a territorial claim or strongly vested interest and which has no traditional treaty rules already governing it, so the framers of the article were able to incorporate much of our accumulated experience with activities in the sea and our knowledge of some of their harmful effects. Second, for a long time to come, it may be expected that most seabed mining and other activities will be carried out by rich, industrialized countries that can afford such environmental protection measures or by the Enterprise, whose activities will be financed in large part by these same countries.

It can be noted, however, that while the scope of the article is broad, the details are sparse and the wording vague, leaving its effectiveness in practice subject to interpretation and enforcement. There is, moreover, little chance that this article will cause *anyone* great inconvenience, since there is unlikely to be any significant commercial seabed mining until well into the next century.

References are made in other articles to pollution resulting from the laying of submarine cables and pipelines, marine scientific research, the use of technologies and the introduction of alien or new species, seabed activities subject to national jurisdiction, and perhaps most importantly, from land-based sources (Art. 207) and from or through the atmosphere (Art. 212), from which sources come over 75 percent of marine pollutants.

Vessel-source pollution falls into several categories. Some results from the normal operations of ships: Oil may be discharged with the bilge water of diesel-driven ships; other pollutants are added by the cleaning of fuel tanks and the discharge of ballast water; nuclear-propelled submarines and surface vessels may release some pollutants; and garbage and sewage are frequently discharged into the sea. By far the largest component of vessel-source pollution, however, is the cargos. Most important, of course, is petroleum and its derivatives, but noxious chemicals, liquid gas and radioactive materials are becoming more important, both in quantities carried by sea and amounts released into the sea. Such releases often occur through deliberate discharge of oil from tankers when they clean their tanks with sea water or pump out ballast water from the oil tanks. Much less frequently but much more dramatically, such releases result from accidents.

Dumping may be considered a form of pollution from ships, but it is generally treated as a separate source of pollution since the ships only carry wastes from the land to be dumped at sea. Such wastes include garbage, sewage sludge, industrial wastes (including many highly toxic pollutants), radioactive matter, dredge spoils and military materials (including obsolete weapons and explosives). As disposal of wastes on land becomes more expensive and difficult, the sea becomes a more attractive disposal site, yet the absorptive capacity of the sea is limited and in some places has apparently been reached.

There is no large body of customary international law covering pollution; as noted earlier, it has become a matter of international concern only recently, and it has generally been dealt with piecemeal through *ad hoc* treaties, beginning only in 1954 and mainly developed through international or regional organizations. The number of such treaties is rather large and they vary considerably in scope, emphasis and effectiveness. Here they can only be summarized briefly.

The huge quantities of petroleum spilled into the sea from sunken and damaged ships during World War II were augmented by more intense pollution resulting from post-war industrialization. By 1954 the situation had become too serious to ignore any longer and a conference in London produced the International Convention for the Prevention of Oil Pollution (OILPOL '54), which entered into force in 1958. For some countries this remains their basic commitment to pollution prevention, since for one reason or another they are not bound by subsequent treaties. It prohibits the intentional discharge of oil and oily mixtures from certain vessels in specified ocean areas, but relies entirely on the flag State (the country in which a ship is registered) to enforce the rules.

IMCO was organized in London in 1958 and in the following year sponsored another international conference on oil pollution in Copenhagen, followed by another in London in 1962. These conferences adopted a number of measures to strengthen OILPOL '54. Following the first two UN Conferences on the Law of the Sea and the *Torrey Canyon* disaster of 1967, IMCO and the shipping and oil industries worked together to deal with the problem. The result was three documents adopted in 1969 that really marked the beginning of a major, sustained effort to deal with marine pollution.

IMCO completed two landmark treaties: the International Convention on Civil Liability for Oil Pollution Damage (the CLU Convention), which effectively doubled shipowners' liability limits and otherwise strengthened civil enforcement procedures; and the International Convention Relating to Intervention on the High Seas in Cases of Oil Pollution Casualties (the Intervention Convention), which for the first time gave *coastal* States the right to take action *outside* their own areas of maritime jurisdiction. Finally, the industry itself produced the Tanker Owner's Voluntary Agreement Concerning Liability for Oil Pollution (TOVALOP), which encouraged tanker owners to clean up oil spills and otherwise covered liability, insurance and compensation. IMCO and the industry separately concluded other agreements in 1971 which, respectively, set up an International Fund for Oil Pollution Damage and a supplement to TOVALOP. Also, regional oil pollution agreements were adopted covering the North Sea (1969) and the Nordic countries (1971).

At this time, international attention turned to pollution problems other than oil spills. There followed in rapid succession a series of agreements that escalate in complexity and importance. These included:

International Convention Relating to Civil Liability in the Field of Marine Carriage of Nuclear Materials, Brussels, 1971.

Convention for the Prevention of Marine Pollution by Dumping from Ships and Aircraft (Oslo Dumping Convention), 1972.

Convention on the Prevention of Marine Pollution by Dumping of Wastes and Other Matter (London Dumping Convention), 1972.

International Convention for the Prevention of Pollution from Ships (MARPOL) 1973.

Convention on the Protection of the Environment Between Denmark, Finland, Norway and Sweden (Stockholm Convention), 1974.

Convention on the Protection of the Marine Environment of the Baltic Sea Area (Helsinki Convention), 1974.

Convention for the Protection of the Mediterranean Against Pollution (Barcelona Convention), 1976.

Kuwait Regional Convention for Cooperation in the Protection of the Marine Environment from Pollution (Kuwait Convention), 1978.

MARPOL Protocol, 1978.[2]

During this period, it may be recalled, the United Nations Conference on the Human Environment was held in Stockholm in 1972 which, among many other important actions, laid the groundwork for the United Nations Environment Programme (UNEP, Nairobi). UNEP's first major, ongoing project was its Regional Seas Action Plans, which began with the Mediterranean and directly led to some of the Agreements listed above. Some of the earlier Conventions were strengthened by amendments and protocols. A long series of marine disasters intensified popular demand for stronger action to protect the sea. And, of course, UNCLOS III deliberated the matter.

Now, for the first time, the United Nations Convention on the Law of the Sea approaches the overall problem of vessel-source pollution and tries to deal with it in a rational manner. It does so mainly through two somewhat contradictory paths.

First, through numerous references to international standards, there is an attempt to limit coastal States' discretion in establishing pollution controls in order to achieve some degree of uniformity in such legislation around the world. Second, the traditional enforcement powers of flag States are now augmented by greater enforcement powers granted to coastal States (those near whose coasts vessels are sailing) and – most radical of all – to port States (those in one of whose ports a particular vessel is). The balancing of interests so important in the Convention is nowhere more evident than in the compromises between the coastal and port States on the one hand and the major maritime States on the other.

Generally speaking, the Convention seeks to attack the problem of pollution by addressing several subproblems. They can be categorized as measures to prevent pollution, reduce and control it, enforce anti-pollution rules, and to assign responsibility and liability for pollution. There are many complex provisions designed to achieve these objectives, many with suitable exceptions and escape clauses, but here these can only be summarized by reviewing the 11 sections into which Part XII is divided.

Section 1 is entitled General Provisions. Three of the five articles have already been discussed but one more (194) deserves some attention. It spells out in some detail countries' general obligation to protect and preserve the marine environment. It contains such sweeping statements as, "States shall take ... all measures consistent with this Convention that are necessary to prevent, reduce and control pollution of the marine environment from any source...." "States shall take all measures necessary to ensure that activities under their jurisdiction or control are so conducted as not to cause damage by pollution to other States and their environments...." "The measures taken pursuant to this part shall deal with all sources of pollution of the marine environment." All of these charges to States, however, are limited in varying degrees by words, phrases or clauses such as "rights," "endeavor," "unjustifiable," etc. Nevertheless, it is an important statement of objectives, and probably expresses as well as anything in the Convention the aspirations of the great majority of people on this planet.

Section 2 mandates or exhorts global and regional cooperation in anti-pollution activities. Following sections provide for technical assistance to and preferential treatment for developing countries and for monitoring and environmental assessment. Sections 5 and 6, entitled International Rules and National Legislation to Prevent, Reduce and Control Pollution of the Marine Environment, and Enforcement, respectively, are the longest, most detailed and most important, and their major provisions have already been discussed. Section 7, Safeguards, modifies the enforcement provisions considerably. Section 8 contains only one article, Ice-covered Areas, which grants coastal States "the right to adopt and enforce non-discriminatory laws and regula-tions" concerning vessel-source pollution within the exclusive economic zone under certain conditions. This is a product of intensive lobbying by Canada, whose Arctic Waters Pollution Prevention Act of 1970, inspired by the voyage of the *Manhattan*, was quite controversial at the time. Section 9 deals with responsibility and liability, 10 with sovereign immunity, and 11 with the continuing obligations of States under other marine environment conventions and agreements, such as those developed by IMO and UNEP.

All in all, the LOSC pollution provisions, with all of their weaknesses and limitations, their ambiguities and restrictions, do represent remarkable and unexpected progress over the near-anarchy that has prevailed in this area. Now all we need is faithful adherence to all of these provisions by all States.

Conservation

In contrast to the waters of the marine environment, the Convention provisions concerning its biota (the plants and animals) are not clustered primarily in one part but are scattered throughout the text. Furthermore, the biota are not considered as integral parts of the environment, but only as "living resources." This term, in fact, dominated the resource-oriented UNCLOS III and now dominates some parts of its chief product. most of the references to the "living resources," which term is not defined at all in the

Convention, are to their allocation among States and measures to maximize their "utilization." Their conservation is considered an economic measure – to assure continued availability of the resource – rather than an ecological measure – to maintain the health of the marine environment.

By far the most important feature of the Convention with respect to the living resources of the sea is the creation of the exclusive economic zone, introduced in Chapter Three, and the most important provisions on fishing are found in Part V, on the EEZ. If all coastal States claim 200-nautical-mile EEZs (and most of them have already), some 90 percent of all of the commercially useful "living resources" of the sea will come under their jurisdiction. This is a radical change from the era before the 1960s, when the only real jurisdiction over fisheries outside the territorial sea was that exercised by some of the intergovernmental fisheries commissions organized on a regional or on a species basis. Many of these commissions still exist but their functions are now considerably reduced.

The notion of an exclusive fishing zone for a coastal State (which later evolved into an EEZ) emerged in modern times in response to a very real problem: overfishing. By the 1960s and early 1970s overfishing had become so severe in such places as the North Sea, the northwest Atlantic and the coastal waters of Peru that some species were virtually exterminated, fishermen were thrown out of work and shore installations related to fishing closed down. In some places the damage was done by local fishermen often outside the territorial sea, beyond the reach of any conservation enforcement officers there may have been.

The real conflicts, however, were between local fishermen and those of the distant-water fleets, the large, modern, well-equipped vessels that often sailed across an ocean to fish. Some, indeed, were factory ships that carried smaller boats which actually caught the fish and brought them back to the mother ships where the catch was immediately canned or frozen and later sold in a convenient market ashore – and not necessarily in its home country. There were bitter disputes and much resentment between, for example, US distant-water fishermen and Peru and Ecuador, between Iceland and boats from West Germany and Britain, and between US inshore fishermen and vessels from Japan, the Soviet Union and Canada, and even from Poland and Spain and other countries.

Since the creation of EEZs, these conflicts have certainly declined, both in number and severity. It is too early to tell, however, whether the fish have benefited from the new regimes. The large factory ships are still operating, but now they do so under license and often with coastal State observers aboard to assure that quotas are not exceeded. Total world catches dropped from their peaks in the 1970s, but there is disagreement over the cause – whether it was economic or biological. Catches have risen again and have exceeded previous records. Partly this has been due to the harvesting of underutilized species; partly to the development of new fisheries, particularly in the Arabian Sea and off the west coast of Africa; and partly to increased harvesting of shoaling pelagic fish needed for fish meal, used chiefly as animal feed in industrial countries. In view of the huge fishery production ever since the adoption of the Convention, one may wonder about the efficacy of the new conservation regimes established in the LOSC for the EEZ and the residual high seas.[3]

In the Seabed Committee and in the earlier stages of UNCLOS III there were three approaches to fisheries questions. First, the developing countries, led by Iceland, Peru and Ecuador and supported later by some developed countries such as Norway, Canada, Australia and New Zealand, advocated expansive fisheries jurisdiction over a broad offshore belt for coastal States. Japan and the Soviet Union, the chief distant-water fishing countries, wanted to retain the status quo, but were willing to concede some preferential rights to coastal fishermen. The United States officially proposed a species approach to fisheries management, but this position was eroded early in the Conference.

The United States was divided on this issue, as on so many others. New England fishermen, complaining of devastating depredation of their traditional fishing grounds by foreign fleets, campaigned hard for a 200-mile EEZ or

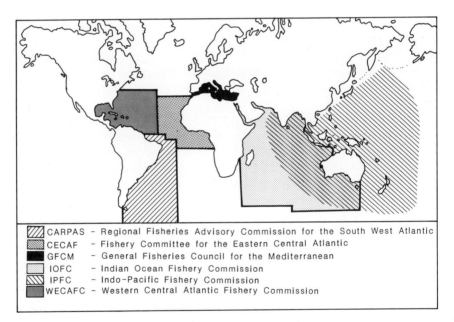

CARPAS – Regional Fisheries Advisory Commission for the South West Atlantic
CECAF – Fishery Committee for the Eastern Central Atlantic
GFCM – General Fisheries Council for the Mediterranean
IOFC – Indian Ocean Fishery Commission
IPFC – Indo-Pacific Fishery Commission
WECAFC – Western Central Atlantic Fishery Commission

Figure 7.3 FAO-sponsored regional marine fisheries bodies. The "competent international organizations" referred to in Article 61 consist of regional fisheries organizations, including those sponsored by FAO. They conduct research, training and educational programs, publish statistics, render technical assistance and otherwise contribute to the conservation of marine living resources. Other intergovernmental commissions sponsored by governments formerly set catch limits, seasons, mesh sizes and so on and enforced their regulations through national courts. Most of them have been dissolved with the widespread adoption of EEZs. (Adapted from FAO 1979, p. 8)

at least an exclusive fishing zone. The shrimpers of Louisiana and Texas, who traditionally fished off the coast of Mexico, were strongly opposed to any such zones, as were the tuna fishermen of Southern California, who were frequently arrested by the authorities of Peru and Ecuador for fishing without licenses in waters these governments considered under their jurisdiction. The fishermen of Alaska and the Pacific Northwest were ambivalent; on the one hand, they resented Japanese and Soviet operations near their coasts, but on the other hand, they themselves often fished far out to sea and off the coast of British Columbia. In the end (near the beginning of UNCLOS III), the New Englanders prevailed, along with other United States interests, and the US in 1974 accepted the principle of a 200-mile EEZ as part of a package deal. Ultimately, even Japan and the Soviet Union had to accept the concept.

Article 61 of the Convention is entitled Conservation of the Living Resources. It requires the coastal State to "determine the total allowable catch of the living resources in its exclusive economic zone," but to do so "taking into account the best scientific evidence available" to ensure against "over-exploitation" of the fishery. There are other provisions concerning maintenance or restoration of harvested species, consideration of species associated with or dependent upon the harvested species, and exchange of fishery data "through competent international organizations." The bulk of Article 62, entitled "Utilization of the Living Resources", details the kinds of laws and regulations that the coastal State may adopt and enforce against foreign fishermen as well as its own.

The next five articles deal with special categories of living resources. Stocks that straddle EEZs of adjacent States or an EEZ and "an area beyond and adjacent to the zone" require shared jurisdiction; highly migratory species (such as tuna, included in a list comprising Annex 1 to the Convention) come under the jurisdiction of "the coastal State and other States whose nationals fish in the region"; marine mammals may be protected more strictly by the coastal State than required by the Convention and all States are required to "co-operate with a view to the conservation of marine mammals...."; "States in whose rivers anadromous stocks [such as salmon] originate shall have the primary interest in and responsibility for such stocks," and fishing of such fish is generally prohibited beyond the EEZ; there is a similar provision for catadromous species [such as some eels]. Most of these provisions deserve brief comments.

The "straddling stocks" issue surfaced late in the Conference and was viewed by many as simply an excuse for coastal States to expand their resource jurisdiction beyond the EEZ. It was resolved in favor of the coastal States probably because the opponents of such broad jurisdiction were exhausted and dispirited by having lost so much already. The United States and some other countries opposed the provisions for highly migratory species and anadromous species because US fishermen could, for example, be prohibited by Article 66(3a) from harvesting tuna even outside Ecuador's EEZ. Article 65, on

Table 7.1   World nominal marine catch (1,000 metric tons)

1970	61 432
1972	62 021
1974	66 339
1976	69 754
1978	70 439
1980	64 383
1982	68 033
1984	73 689
1985	75 138
1986	80 345

Source: *FAO Yearbooks of Fishery Statistics*

marine mammals, is much stronger than it was in earlier versions, and much credit for this achievement must be given to the Connecticut Cetacean Society (now called the Cetacean Society International), one of a number of non-governmental organizations (NGOs) that were active and influential at UNCLOS III.

Articles 87 and 116–20 provide for the conservation and management of the living resources of the high seas and differ little from the 1958 Geneva Convention on the same subject. The provisions are similar to those for Conservation in the EEZ, but they are less detailed, there is less incentive to cooperate there than in the EEZ and there are no enforcement provisions. This is indicative of the fact that on the high seas – a commons – there is still little immediate reward for any State to regulate fishing by its nationals; there is clearly such an incentive within EEZs.

These weaknesses may not be immediately important since, with the transfer of the best fisheries from the high seas to the EEZs, there can be expected relatively little high seas fishing for a while. As the search for new fisheries and non-traditional species intensifies, however, in response to rapidly increasing demands for fisheries products, the conservation ethic of high seas fishermen will be severely tested far beyond the reach of fishery enforcement vessels and aircraft.

In both its EEZ and high seas parts, the negotiators of the Convention chose not to adopt the modern concept of optimum sustainable yield of fisheries to allow for inadequacies in our understanding of fisheries biology and for normal fluctuations in the size of fish stocks, and instead enjoined States, in determining their allowable catch, to take measures "to maintain or restore populations of harvested species at levels which can produce the *maximum sustainable yield* ..." (Arts 61(3) and 119(1) (1); emphasis supplied). Notwithstanding this adherence to the goal of production rather than conservation, States *are* urged to use "the best scientific evidence available ..." to conserve the species being exploited.

Other references to conservation of living resources may be found in the sections of the LOSC covering the territorial sea, straits, archipelagic waters, enclosed and semi-enclosed seas, the Area, pollution and transfer of technology. It is clear, then, that the concept has become widely accepted. How it will be interpreted and applied, however, remains to be seen. Like the provisions on pollution, it may be honored more in the breach than in the observance. So far, it appears that most States are, in fact, honoring at least the spirit of the LOSC provisions on both pollution and conservation. We can observe the cleansing of some bodies of water (notably the Baltic Sea) and the return of some overfished species (notably the Pacific halibut). If these trends continue, we may be able to maintain to a reasonable degree the health of the global sea. But we cannot be certain until all States cooperate to treat the sea as truly "an indivisible ecological whole."

8 *Marine scientific research*

Marine scientific research? What is geographic about that? Or political? Why shjould it be included in a book on the political geography of the sea? These questions are fair ones and not difficult to answer.

We began this book by stating that it is a book about the sea. Some of what is known about the sea has been learned very gradually, in tiny increments, over scores of generations, by thousands of anonymous fishermen, navigators and adventurers. Most of what is known now about the sea, however, especially about its deeper and remoter portions, has been learned during the past century through well-organized expeditions of scientists, well-trained in a number of disciplines and equipped with the best available technology. In the process of exploring, studying and publicizing the geology of the ocean basins, the nature of the sea water, movements of water in the ocean basins, plant and animal life in the sea and the interactions among the sea and the atmosphere and the land, they have steadily developed a more complete and accurate picture of the physical geography of the sea.

Without this knowledge, there could be no real economic geography of the sea, no systematic utilization of the resources of the sea for the benefit of man. There would still be some deep-sea fishing and some offshore collection of shellfish, finfish, seaweeds, sand and gravel, and some other useful things found on the shallower parts of the continental shelf and in the waters above it, but certainly no massive distant-water fishing fleets, no large-scale offshore oil and gas production, no deep seabed mining, or harnessing of the sea itself for energy. Without this extensive and intensive utilization of the sea, there would be little need for zones and boundaries, for rules and procedures to regulate such uses. Admiralty law would serve the needs of the shipping industry and traditional international law would suffice for a narrow territorial sea and for the high seas. Because science has led us to complex and often conflicting uses of the sea, the political geography of the sea has developed to help us understand the problems and try to work out rational and peaceful solutions to them.

But what is political about marine scientific research? A generation ago, nothing. At that time, a handful of government agencies and private institutions conducted limited research in scattered, relatively small areas of

Figure 8.1 Marine scientific research. Most of the research into the marine environment is slow, undramatic, often tedious, sometimes frustrating but occasionally supremely rewarding when some major discovery is made or confirmed. Mostly, however, as illustrated by this photograph of experiments being conducted on the seabed near Grand Bahama Island, it consists of the collection and analysis of innumerable bits of data gathered over a period of time. (United Nations/D. Clarke)

the sea, generally not far from shore. The researchers were nearly all from a few rich countries; they functioned, in fact, much like a fraternity of scientists united in the age-old quest for knowledge. Rarely was permission for such research needed or sought, for most of the work was done in home waters or beyond even the six-mile territorial seas of the other States, in international waters. The peoples in the colonies, for the most part, had little interest in science or in the sea, and few opportunities to pursue what interest they might have had. The results of the research were generally published in scientific journals read by few outside the scientific community and specialized government agencies. This casual, even adventurous, era of marine science has faded and nearly vanished since World War II, and especially since the late 1960s.

As we have already noted, the process of decolonization since World War II has had a profound influence of the development of international law in general and on the Law of the Sea in particular. There was considerable suspicion and even some hostility towards an international system created, in large measure, by the imperialist countries to serve their own purposes with little regard for the subject peoples. Now the newly independent States looked anew at the law and custom of marine scientific research, for example. The Cold War, the frictions of decolonization, onrushing technology development, increasing demands for resources of all kinds, fierce competition in the international marketplace, and a general and bewildering complexity characterized the world into which they had emerged.

Under the circumstances, it is not surprising that they tended to view with suspicion and resentment, not entirely unjustified, strange-looking foreign vessels skulking about off their coasts doing mysterious things with weird equipment. These ships could well be preparing either for rapacious plundering of offshore resources or a military operation of some kind. Lacking, in most cases, both a scientific and a maritime tradition, new countries tended to try to restrict marine scientific research in a variety of ways. UNCLOS III provided an ideal forum within which they could attempt to do this in an organized and concerted manner.

Now the political dimension is exposed. In UNCLOS III there was at first a sharp division on the issue of research. Although there were many points of disagreement, the basic issue was whether coastal State consent was necessary before any specific research project might begin in the State's EEZ or on its continental shelf, and what conditions, if any, the coastal State might impose on such research if consent is granted. The "North–South" confrontation on this issue posed many problems in the Third Committee, that which had been assigned what were supposedly the least important and least controversial topics.

The issue at UNCLOS III

There are certain requirements that are essential to useful marine scientific research (MSR). These cannot be discussed in detail here, but a list would be helpful in understanding the debate.

(a) *Access to data.* This means first the collection of data and then its interpretation. In most cases, in marine science, this means field work as well as library and laboratory work. It also means the exchange of both raw data and interpretive literature within the worldwide scientific community with as few restrictions as possible.

(b) *Access to ocean areas of scientific interest, freedom of movement and access to ports.* By their very nature, oceanography and its related disciplines are mobile, studying and comparing different portions of the global sea or tracing the migration patterns of marine creatures or measuring the flows of currents at various depths. Research voyages are normally planned two to three years in advance, but often last-minute changes are required, so flexibility is essential. Ships depend on ports, and any limitations on access to them can seriously reduce the scientific value of a cruise and add to its cost.

(c) *International cooperation.* Because a particular research project, to be effective, may involve work in the waters off more than one country, coordination through an international agency, or at least between governments, is esssential. Since no country has a monopoly on scientific knowledge or talent, international cooperation is necessary for scientists of all nationalities to benefit from one another's knowledge and experience. Modern large-scale and sophisticated expeditions often require expensive equipment and logistical suport which may only be provided by international cost-sharing.

Like nomads on land, free-ranging research vessels that wish to cross boundaries at will pose problems for governments in general and especially those still not entirely secure about their newly acquired sovereignty. The new countries have also come to appreciate the value of science and tend to resent being excluded from the scientific community and from the fruits of its work. Thus, at UNCLOS III many in the Group of 77 strove to accomplish two contradictory objectives at once: to restrict or at least control scientific research within the marine areas under their jurisdiction, and to participate more actively in such research and share in its benefits. This was one more area in which the developing countries tried to advance their movement for a new international economic order.

Broadly speaking, then, the creation of broad new areas of coastal State jurisdiction at sea has led to a problem that scarcely existed until very recently. Prior to 1973, 50 percent of all US MSR (marine scientific research) was conducted on or over the continental shelf and another 30 percent within 200 nautical miles of the coast. Most other countries with less research capacity conducted a much larger proportion of their MSR in these areas. Worldwide, well over 80 percent of MSR has traditionally been done in areas that may now fall under national jurisdiction. In UNCLOS III the customary freedom of research, as it existed even under the 1958 Geneva Convention on the High Seas, was challenged by the developing coastal States.

There was at UNCLOS III no manifest objection to marine scientific research as such. Attempts by many developing coastal States to restrict and control it tended to revolve around three basic demands, all of which were at odds with the basic requirements of such research that we described above.

(a)   A distinction should be made between "pure" and "applied" research, with far more restrictions to be placed on the latter because of its commercial and military potential.

This issue evaporated during the latter stages of the Conference as the developing countries came to understand that this is a distinction without a difference. The results of even the "purest" research, which is undertaken to satisfy purely intellectual curiosity (if such a thing exists anymore), can quickly be utilized for commercial and/or military purposes, often in unpredictable ways.

(b)   MSR should not be allowed to proceed within the limits of national jurisdiction without the explicit advance consent of the coastal State.

The scientists and their representatives at the Conference argued, however, that for a number of reasons, waiting for such consent and the risk of its denial could diminish the success of an expedition or make it infeasible altogether. Ultimately, this latter view was tacitly accepted by the developing countries.

(c)   Scientists from the coastal State must participate in every phase of the research from initial planning through execution to analysis and dissemination of the data collected.

While the scientific community did not object in principle to this demand, they did point out a number of practical difficulties, any of which could present real problems for the success of a mission.

As was true in nearly every segment of the sprawling and seemingly interminable UNCLOS III, the long, sometimes tedious, sometimes heated debates on MSR slowly produced a meeting of minds of the delegations, or at least a reduction of the distance between their positions. Part XIII of the United Nations Convention on the Law of the Sea is by far the most comprehensive and detailed legal statement on marine scientific research ever produced. Its details are not of interest to a political geographer but its broad outlines are.

The new MSR regime

The first four articles (Arts 238–41) lay out general provisions for MSR. All States and competent international organizations have the right to conduct it "subject to the rights and duties of other States . . ."; they shall promote and facilitate it; it shall be conducted exclusively for peaceful purposes and in a manner compatible with other provisions of the Convention; and it "shall not constitute the legal basis for any claim to any part of the marine environment or its resources." The next three articles spell out the requirement for and means of achieving international cooperation in marine scientific research, including the "publication and dissemination of information and knowledge."

Section 3, containing 13 articles, is the heart of the whole part. Among other things, it spells out in some detail the conditions under which MSR in the EEZ and on the continental shelf "shall be conducted with the consent of the coastal State." Although Article 246(3) states, "Coastal States shall, in normal circumstances, grant their consent . . .", such consent may be withheld for a variety of reasons, and, of course, notwithstanding the general requirement of consent, it could be delayed indefinitely. The key article, the compromise that finally broke the stalemate over consent, is Article 252, Implied Consent. It provides that "States or competent international organizations may proceed with a marine scientific research project six months after the date upon which the information required pursuant to Article 248 [about the planned project] was provided to the coastal State" unless within four months the coastal State has denied the request or responded to it in other ways that do not mean consent. In other words, silence implies consent.

Other articles provide for the participation of the coastal State in the research, dissemination of research results, suspension of the MSR activities under certain circumstances, MSR in the Area under a separate set of rules, and a variety of other conditions. All in all, under the new regime, marine scientific research can still be conducted within the new zones of national jurisdiction, but with less certainty and at greater cost than before UNCLOS III.

The United States and a few other industrialized, maritime countries have expressed their displeasure with the LOSC rules on MSR. They contend that they are unduly restrictive and, if followed too strictly or if interpreted too narrowly, they could impede legitimate MSR and thus deny everyone its potential benefits. The Convention rules are certainly more restrictive than marine scientists would like and the objections of the US have some validity. However, it is very likely that the education process evident at UNCLOS III is still continuing.

Most countries now understand the value to all mankind of more complete and accurate information about the sea and are becoming more cooperative with researchers. Many, moreover, are getting into the game themselves, albeit in some cases rather slowly and timidly. India, Brazil and Indonesia as well as a few other developing countries, however, are energetic and bold in this area and are beginning to catch up with the developed countries. It seems likely that, in general, discord over MSR will diminish and cooperation increase in the near future. Either success or failure in reaching full international cooperation in marine scientific research, however, will provide a first-rate topic for politico-geographic analysis.

9 Additional themes and topics in the Convention

Because of the nature of the evolution of the Law of the Sea, and the manner of organization of UNCLOS III, the United Nations Convention on the Law of the Sea does not lend itself to analysis of all topics of interest to a political geographer in an orderly, systematic fashion. Thus, for example, if one is interested in fishing, one must find references to it in dozens of articles scattered through the main text and three annexes. Similarly, some topics of great interest to political geographers are scarcely mentioned in the Convention or omitted altogether. Some of the topics omitted will be discussed in subsequent chapters, but here we will include some that deserve more attention than they received in the Convention. The selection is somewhat arbitrary, of course, and is not meant to be definitive. A careful reading of the Convention and its appendages will reveal many more topics worth the scrutiny of a political geographer.

Assistance to developing countries

It was pointed out in Chapter Two that UNCLOS III was viewed in large measure by the Group of 77 (the overwhelming majority of the countries of the world) as an important vehicle for helping to develop a new international economic order, to redress the gross imbalance in wealth and power so characteristic of the world for centuries. At the same time, the rich, industrialized, urbanized, imperialist powers (including the Soviet Union and its allies), which were also for the most part the major maritime powers, saw as their task to "hold the line", to resist the expansion of national jurisdiction out to sea, to nullify or at least emasculate the common heritage principle, to retain as much as possible of the traditional Law of the Sea.

This is a great oversimplification, of course; only rarely were there head-on North–South confrontations, and the socialist countries often took a middle position. There was, despite the underlying tension, considerable compromise and cooperation. Part of this was due to a genuine determination to produce a generally acceptable treaty that would prevent the sea from becoming a gigantic arena of international conflict. This objective was achieved by constant concessions by all sides. One of the greatest compromises of all has scarcely been mentioned in the literature – the many concessions made to the special needs and interests of the developing countries.

The Preamble of the 1982 Convention deserves much more attention than it has received. "The States Parties to this Convention," it begins,

> Bearing in mind that the achievement of these goals [laid out in the first four paragraphs] will contribute to the realization of a just and equitable international economic order which takes into account the interests and needs of mankind as a whole and, in particular, the special interests and needs of developing countries, whether coastal or land-locked . . .

Thereafter, there are numerous provisions in the Treaty itself – and in one of the annexes and one of the resolutions that form part of the total Convention package – that specify how these interests and needs are to be accommodated. One whole part, for example (Part XIV, Development and Transfer of Marine Technology) is devoted exclusively to this objective. Paragraph 2 of Article 266, the lead article in this part, deserves citing *verbatim*:

> States shall promote the development of the marine scientific and technological capacity of States which may need and request technical assistance in this field, particularly developing States, including land-locked and geographically disadvantaged States, with regard to the exploration, exploitation, conservation and management of marine resources, the protection and preservation of the marine environment, marine scientific research and other activities in the marine environment compatible with this Convention, with a view to accelerating the social and economic development of the developing States.

The remaining dozen articles in this part, some quite detailed, suggest or specify many ways of carrying out this objective. Annex III, Basic Conditions of Prospecting, Exploration and Exploitation [in the International Seabed Area], devotes considerable attention to this objective as well, and Resolution II, governing Preparatory Investment in Pioneer Activities Relating to Polymetallic Nodules, also requires the transfer of technology to the Authority and "training at all levels for personnel designated by the Commission" of the Authority.

Other articles require or refer to technical assistance or training in connection with fishing, seabed mining, pollution and marine scientific research. Generally, these provisions were adopted with little real controversy. Those on MSR, however, and especially those on seabed mining and the organization and operations of the International Seabed Authority, were strongly, even bitterly, resisted by the major industrial powers, and especially by the USA. This was one of the major reasons advanced publicly by President Reagan for rejecting the entire Convention.

How important this really is to the United States, whether the Convention provisions will be modified in practice, whether they really make any difference to the developing countries, are all questions that cannot be answered at the moment. What is clear now, however, is that if the relevant provisions of the Convention, even excluding those relating to MSR and seabed mining, are carried out faithfully, most developing countries will gain a great deal from them. Whether these gains will be cancelled out by losses sustained through other provisions of the new Law of the Sea, however, remains to be seen.

Navigation

Despite dramatic advances in transportation technology during the 20th century, most international trade still moves by sea for at least part of its journey from origin to destination; well over 80 percent of it, in fact. And shipping is still – along with fishing – one of the two most important uses of the sea, despite advances in technology that have enabled us to use the sea in many more ways. But shipping itself is governed largely by admiralty law. The Law of the Sea – and the political geography of the sea – are concerned primarily with freedom of navigation.

ADMIRALTY LAW

The Law of the Sea is sometimes confused with maritime law, or the law of admiralty. Admiralty law comprises the most important part of private law that deals with the shipping industry. Its long history and separate international traditions, as well as its link with a single industry, distinguish it from other branches of law. It includes a large body of rules, concepts and legal practices governing such central concerns of the industry as marine insurance, carriage of goods under bills of lading, charter parties, rights of seamen and maritime workers, collision, salvage, maritime liens and ship mortgages, and limitation of liability.

Admiralty law and the Law of the Sea, a branch of public international law, overlap somewhat in such matters as the pollution of the sea by ships, the international rules of the road, communications at sea and from the sea, the safety of life at sea, procedures in distress situations, and similar areas. In the Law of the Sea these items are included in the provisions for the regime of the high seas, the portion of the sea beyond the jurisdictional waters of coastal States. Like all other aspects of international law, however, and like admiralty law, these rules are enforced primarily in the courts of individual countries.

As recounted earlier, the major maritime powers – the USA, USSR, the UK, Norway, France, Japan and so on – staked out as one of their chief goals, perhaps the single most important one at UNCLOS III, the retention of the maximum possible freedom of navigation in the face of expanding national jurisdiction at sea and the explosive growth of both new and traditional competing uses of the sea. Many coastal States, however, especially developing countries with no navies or merchant fleets, saw freedom of navigation as simply a euphemism for domination of the sea by a few rich and powerful countries, an instrument of old-fashioned imperialism and modern neocolonialism. It was simply irrelevant to their own interests or perhaps even an active threat to them. In their scale of priorities, freedom of navigation ranked rather low and the maritime States had to bargain hard to preserve it.

As a result, freedom of navigation, or shipping, is not separated out for special treatment in a separate part of the LOSC, but rather is handled in a number of articles under several other headings. The basic principle is retained in Part VII, High Seas, wherein Article 81, titled Freedom of the High Seas, states: "The high seas are open to all States, whether coastal or land-locked". But Article 87 lists "freedom of navigation" as only one of six freedoms comprising the freedom of the high seas. Other articles in this part, nine of them, deal specifically with shipping. Among other things, they provide that: "Every State, whether coastal or land-locked, has the right to sail ships flying its flag on the high seas" (Art. 90), and "Warships on the high seas have complete immunity from the jurisdiction of any State other than the flag State" (Art. 95).

One of the most interesting and politically significant of these articles is that (Art. 91) covering the nationality of ships. After stating that ships "have the nationality of the State whose flag they are entitled to fly," it goes on: "There must exist a genuine link between the State and the ship." This is an attempt to limit, but not outlaw, the so-called "flags of convenience," which are officially called "open registry" and which shipowners and some governments consider "flags of necessity." The "reflagging" of Kuwaiti tankers operating in the Persian Gulf (placing US flags on them and granting them US naval protection) during the Iran–Iraq War of the 1980s introduced a new political element into what had been at least in large measure an economic arrangement. The "genuine link" concept remains controversial in theory and difficult to administer, but it may lead to a structural change in the world shipping industry.

Other navigation-related provisions include those concerning sea lanes and traffic separation schemes in the territorial sea, in straits and in archipelagic waters; and a number that are designed to prevent conflicts between navigation and other uses of the sea. Finally, some of the pollution articles, especially as they apply in the newly expanded areas of national jurisdiction, may, by permitting coastal States to establish their own requirements for ship construction, operation and routing, raise shipping costs and affect service. It will be some time, however, before it is possible to judge accurately whether the navigation provisions of the LOSC represent a net gain or a net loss for the international community. The key test is likely to be the interpretation and application of the principle of high seas freedom of navigation within the exclusive economic zone. Here is where law, politics, economics, custom, security and national sensibilities are most likely to generate frictions.

Aerial navigation

The traditional Law of the Sea was virtually silent about aircraft, and international aerial navigation is regulated chiefly by ICAO, the United Nations International Civil Aviation Organization, headquartered in Montreal. Now the situation is different. The role of ICAO has not changed, but now the Law of the Sea also includes a number of provisions pertaining to aircraft. This was necessitated by the expansion of coastal State maritime jurisdiction and by the insistence of the major powers on maximum freedom of navigation in the air as well as on the sea.

Thus, while there is still no right of innocent passage for aircraft over the territorial sea, the new rules for transit passage of ships through straits used for international navigation and through archipelagic sea lanes apply to aircraft as well. Other provisions of the 1982 Convention pertaining to the exclusive economic zone, the high seas and pollution apply, with appropriate modifications, to aircraft as well as ships. Again, these provisions represent a balancing of interests, with the most important provision probably being that won by the major powers over the vigorous objections of some straits States and others: the right of transit passage over straits. All in all, international air transport is well served by the new Law of the Sea.

Artificial islands, structures and installations

Under the 1982 Law of the Sea Convention, States retain their traditional right to lay pipelines and cables on the floor of the sea, even within other States' new zones or jurisdiction, though with some modifications, mostly involving coastal State consent in planning. The matter of artificial islands, structures and installations, however, is different. The only 1958 rule that relates to it is one article in the Continental Shelf Convention that provides that installations on the shelf have no territorial sea of their own, nor may they affect in any way the delimitation of the territorial sea. Now the situation is more complex.

First of all, since 1958 there has been immense growth in the offshore oil industry around the world, and with it thousands of new oil wells, platforms and other structures placed on the continental shelf. That alone would have required more elaborate rules than previously, but there are in addition many other kinds of offshore islands, structures and installations in use now, projected for construction very soon, and contemplated for the intermediate future. These include such things as offshore oil tanker terminals, scientific research apparatus, artificial reefs to shelter fish and other marine creatures, a planned underwater hotel and many more, ranging all the way to offshore airports and even whole cities. Clearly, new rules are necessary to control or at least regulate them.

Secondly, the LOSC uses the three separate terms (islands, structures and installations), sometimes together and in various combinations, but nowhere are any of these terms defined. One or more of them appears in nine articles dealing with the EEZ, the continental shelf, the Area, pollution and marine scientific research.

We may recall the observation made in Chapter Five about the difficulty of identifying an island that is "naturally formed." With the likely development of many artificial islands which, like artificial structures and installations, may not have territorial seas, EEZs or continental shelves of their own, nor affect the delimitation of these zones, we may be absolutely certain that considerable human ingenuity will be devoted to the creation of islands that could possibly pass as "naturally formed" so as to be able to claim such zones. We may also anticipate that, despite provisions in the Convention for placement of artificial islands *et cetera* where they will not obstruct major shipping lanes or otherwise interfere with other legitimate and important uses of the sea, and for safety zones up to 500 meters broad to be established around them, there will be accidents involving them that may not always be handled through the dispute settlement procedures elaborated in the Convention.[1]

Furthermore, the exclusive right of the coastal State to construct them in its EEZ and to authorize and regulate their construction, operation and use is limited by certain duties, including removing them to ensure safety of navigation when they are no longer in use. With scores, perhaps hundreds, of offshore oil wells around the world now reaching the end of their useful life, the strict observance of this duty may present some problems that are not easy to solve. It may not be long before the continental shelves of the world are as littered with junk as the exposed land.

Slavery, piracy, unauthorized broadcasting, narcotraffic

There are some exceptions to the exclusive jurisdiction of a flag State over its ships on the high seas and they are spelled out in Articles 99–110 of the 1982 LOSC. In each case, all States are not only permitted but required to cooperate and/or act individually to the fullest extent possible to suppress these illegal activities. Though the first of these, the transport of slaves, is fortunately nearly extinct, the other three are still of concern. Until recently they could be considered simply as criminal activities and outside the scope of political geography. Now, however, since they are so frequently connected with political activity in many countries, including "national liberation movements" and irredentist movements, and linked to gun running for political purposes, they are of interest to political geographers.

Slavery and piracy were included in the 1958 High Seas Convention; the other two activities were not. The 1982 definition of piracy is more elaborate than that of 1958 and now includes aircraft, but it is still inadequate from a legal standpoint. Hijacking, for example, is not considered piracy since only one vessel – the victim – is involved. Thus, ships were not required under the Law of the Sea to come to the aid of the victims in either the *Santa Maria* (1961) or *Achille Lauro* (1985) cases. In fact, in no recent case of illegal activity on the high seas in which perpetrators have been brought to justice have they been charged with piracy. Similarly, Article 108, which requires cooperation "in the suppresssion of illicit traffic in narcotic drugs and psychotropic substances engaged in by ships on the high seas contrary to international conventions," has not yet been effectively applied. So-called "pirate broadcasting" by unauthorized persons from ships outside the territorial sea has been largely suppressed, but generally by European States acting alone or in concert, and long before UNCLOS III. The new article 109 of the Convention remains to be tested.

Marine parks, reserves and sanctuaries

One of the encouraging features of marine affairs today is the rapidly spreading movement to designate specific offshore areas for special protection because of their ecological, scenic, historical, archaeological or scientific value. These areas are known in various countries as marine parks, marine reserves or marine sanctuaries. There is nothing in the Law of the Sea that specifies the establishment of such protected areas; they are established entirely by national legislation in areas entirely under the jurisdiction of the coastal countries. Nevertheless, they are encouraged by the LOSC requirements that States take affirmative action to preserve and protect the marine environment. The movement began with the establishment of two marine sanctuaries by the United States in 1975. Since then the United States has added more such areas and many other countries have done the same.

The nearest approach to this kind of protection that appears in the Law of the Sea Convention is two new articles on archaeological and historical objects. One of the General Provisions found near the end of the Convention is Article 303, which says: "States have the duty to protect objects of an archaeological and historical nature found at sea and shall co-operate for that purpose." It goes on to refer to Article 33, on the contiguous zone, as being useful in controlling traffic in such objects. Article 149 reads: "All objects of an

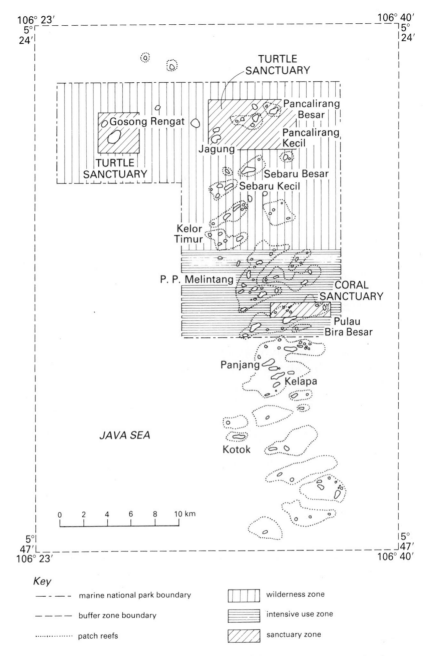

Figure 9.1 Marine sanctuaries in Indonesia. One of the more encouraging developments in marine affairs since 1982 is the rapidly spreading movement among developing countries to establish marine parks, reserves and sanctuaries to protect underwater areas of special scientific, historical, cultural or aesthetic interest. As usual, however, this use of the sea often conflicts with other uses, most commonly navigation and resource development, just like similar conflicts on land. It may be difficult to resolve these conflicts in favor of long-term environmental and cultural considerations, but individuals and non-governmental organizations can be influential in this effort. (White 1984, p. 183)

archaeological and historical nature found in the Area shall be preserved or disposed of for the benefit of mankind as a whole . . ." It is certainly good that these provisions are in the Convention, since even with their limitations and restrictions, they do lend the stamp of international concern to this very important matter. They do not, however, cover the sites in which such objects are found, only the objects themselves, and this broader protection is left entirely to individual States. Neither is there any method specified for protecting these objects, nor any specific enforcement mechanism, or even a definition of the term "objects of archaeological and historical interest."

It has been suggested, primarily by ecology- and history-minded groups in the USA, that the *Titanic* and its resting place be declared an international marine park. Regrettably, there is simply no mechanism at present for doing so, even leaving aside the very real questions of definitions, jurisdiction, salvor's rights under admiralty law and so on. Perhaps some day the international community will provide for such international protection in the sea beyond the limits of national jurisdiction, but meanwhile we can only applaud and encourage more national legislation of this kind. Even here, though, problems arise with regard to the degree of protection provided, multiple use, public access, costs, enforcement and other problems found in parks, reserves and sanctuaries on land.

Parties to the Convention

Finally, Article 305 provides that, for the first time in history, entities other than States are permitted to become parties to an international treaty of such scope and importance. (The United Nations Charter, which was signed by India and the Philippines before they became independent and by the Ukraine and Byelorussia though they are still far from independent, offers some precedent.) Among the entities listed are:

- Namibia, represented by the United Nations Council for Namibia;
- Three kinds of self-governing territories and associated States which "have competence over the matters governed by this Convention, including the competence to enter into treaties in respect of those matters;"
- "International organizations, in accordance with Annex IX."

Both the UN Council for Namibia and the Cook Islands signed the Convention at Montego Bay. Other such non-independent territories may accede eventually. This example of limited sovereignty should prove of interest to political geographers as well as international lawyers and is worth additional investigation.

The first two of these categories engendered little opposition at UNCLOS III;

the third generated one of the Conference's longest-running controversies. The ultimate compromise on this question was Annex IX, which consists of eight articles, some of them very complex indeed. For our purposes, only Article 1 is important. It reads:

> For the purposes of article 305 and of this Annex, "international organization" means an intergovernmental organization constituted by States to which its member States have transferred competence over matters governed by this Convention, including the competence to enter into treaties in respect of those matters.

In practice, this means the European Community, at least for the present.

So far the European Community has adopted common policies on shipping, marine pollution and, most important and most difficult to achieve, fishing. It has also achieved a large degree of cooperation among its members in such relevant matters as boundary delimitation, offshore oil and deepsea mining, and in coordination of policies at UNCLOS III. The common fisheries policy was painfully negotiated between 1966 and December 1981. Final agreement came just in time to provide the conclusive evidence to the Conference that the EC member countries really had transferred competence over important marine matters to the intergovernmental organization and thus qualified it to sign the Convention in its own right. Most of the member States also signed it, although the UK and the Federal Republic of Germany did not. This common fisheries policy is so important that, notwithstanding space limitations, it does deserve a brief review here and additional comments in Chapter 12.

The members have ceded to the Community competence over all species of fish, whether aquatic or marine, and all crustaceans and molluscs, but not corals, sponges or shells other than molluscs. This includes marketing and trade in these products and the fats and oils of marine mammals, as well as the harvesting of these animals exclusively for the purpose of catching such animals. The geographical scope of the policy is uncertain, since maritime boundaries are still within the province of the individual member States, and many of them are disputed, including some in the Americas. This policy and regulations promulgated to implement it are applicable to individual fishermen and vessels engaged in fishing in the areas in which the policy applies. As might be expected in such a complex and important pioneering venture, many details remain to be worked out and undoubtedly mistakes will be made. Nevertheless, the EC is blazing a trail that may eventually be followed by other intergovernmental organizations around the world.

10 *The polar seas*

The Third United Nations Conference on the Law of the Sea did not consider a number of issues relating to the sea, and thus they are barely mentioned in the Convention it produced or ignored altogether. Among these are non-nodule resources of the International Seabed Area, use of the sea for the production of energy, shipping, use of the airspace over the sea, the polar regions and military uses of the sea. There are various reasons why these issues (and others) were omitted from the UNCLOS III deliberations, but the reasons do not concern us here. It should be made clear though, that all of them are important politicogeographic matters, and any or all of them may gradually develop into contentious issues or burst suddenly into public consciousness as a result of some dramatic incident. Of these, the two that require consideration here are the polar regions and military uses of the sea, to each of which we devote a chapter.

The Arctic

The only reference in the Law of the Sea Convention to any of the special conditions that prevail in the polar seas is Article 234, on ice-covered areas, which was discussed in Chapter 7 since it deals exclusively with pollution prevention legislation. Even this article carefully avoids the basic problem in the polar seas: the status of ice in international law. Is ice merely water in a solid state and therefore subject to all the rules that apply to the sea? Or is it merely a rather unusual form of land and therefore subject to the same rules that apply to land territory? Do States have the right to claim ice-covered portions of the sea on the same grounds as they claim land territory, or must they distinguish between permanent ice formations (such as shelves) that are permanently attached to the land and temporary formations such as bergs, islands and floes? Are ice islands islands or merely part of the sea, or, when used as bases for exploratory expeditions or scientific research, are they akin to vessels on the high seas? These and other questions about ice are not abstractions; they are of very real importance in determining national jurisdiction.

An even more important question relating to jurisdiction is raised by the so-called "sector theory." Canada has since 1907 claimed sovereignty over all territory to the north of its continental land mass; i.e., between 141°W and 60°W, all the way to the North Pole. However, successive Canadian governments have generally indicated that this claim of sovereignty extends only to the land within this sector, not to the ice or open water beyond the territorial sea. But more recently they have claimed as internal waters all of the sea within the Canadian archipelago, and in 1987 made this claim formal by drawing closing lines connecting the outermost points of the outermost islands. This new claim has not been recognized by any other States and has been directly challenged by the USA. Neither does any other country bordering the Arctic Ocean formally recognize the sector theory as a valid basis for a claim to territory. The Canadians continue to insist that because the Arctic Ocean is nearly all ice-covered, it cannot be considered high seas and must be subjected to a special legal regime. This position has similarly attracted little support.

The position of the Soviet Union with regard to these matters had been most ambiguous until quite recently. Technically, it rejected the sector theory, as noted above, yet both the government and some Soviet legal experts at times did refer to it vis-à-vis the waters to the north of the mainland. At times the government designated some straits and bays as "historic waters" and excluded foreign vessels from them, but at other times it permitted innocent passage even of foreign warships through these same waters. It had required prior authorization for entry of foreign warships into Soviet internal waters and ports, but had not published charts showing the baselines enclosing the internal waters.

Since signing the United Nations Convention on the Law of the Sea in December 1982, however, the Soviet Union has been bringing its policies and domestic legislation into conformity with its provisions. Among these actions are the following:

- The Law of the USSR State Boundary of 1982 establishes a 12-mile territorial sea and provides for straight baselines along the coast. It also, however, claims as internal waters "the waters of bays, inlets, coves, estuaries, seas and straits, historically belonging to the USSR" without naming them.
- Amendments to the 1968 Edict on the Continental Shelf of the USSR implement some relevant provisions of the LOSC.
- Legislation in 1984 established an EEZ of 200 miles along all Soviet coasts.
- Rules recognizing the right of innocent passage of foreign warships through Soviet territorial waters were confirmed in 1983.
- Decrees in 1984 and 1985 fixed the coordinates of straight baselines along all Soviet coasts.
- Several decrees and statutes have been adopted pertaining to protection of

the marine environment and establishing procedures for the conduct of marine scientific research in waters under Soviet jurisdiction.

It would appear that the Soviets are no longer as sensitive about the security of their Arctic seacoast as they were when their navy was weak. They also seem satisfied that the LOSC, on the whole, protects Soviet economic, environmental, resource, political, cultural and strategic interests in the Arctic. This does not mean, however, that Soviet maritime policies are now fixed and uniformly administered in conformity with the Convention. The Black Sea incident of March 1986, when Soviet warships challenged two US warships near the Crimea, seems to indicate that other considerations may at times supersede their compliance with the Convention and perhaps even with their own legislation.[1]

Another question unique to the Arctic arises from the relationship between the Law of the Sea, and in particular the continental shelf doctrine, and the Svalbard Treaty of 1920 (which became effective in 1924). This treaty placed the Svalbard archipelago, including its main island of Spitzbergen, under the "full and absolute sovereignty" of Norway, but also granted to citizens of parties to the treaty, including the Soviet Union, equal rights to exploit the resources of the islands and their territorial waters. The only treaty power currently taking advantage of this provision (aside from Norway) is the Soviet Union, which mines coal on Spitzbergen. The question is whether the treaty provisions extend to Svalbard's EEZ and continental shelf or whether some other arrangements must be made for allocating resource jurisdiction beyond the territorial sea. In any case, there are complicating factors, such as the lack of any break in the continental shelf between Svalbard and mainland Norway, the uncertain maritime boundary between Norway and the USSR, and the need to define a maritime boundary between Svalbard and Denmark's Greenland, soon to become independent.

The Arctic is also of special interest to the scientific community and has been for generations. There has generally been good cooperation among scientists working in this hostile region, and research data have generally been shared. Since World War II, however, the Arctic has become one of the most intensively militarized regions in the world. The two superpowers face each other across the pole and share a very narrow passage between the Diomede Islands in Bering Strait. Their submarines regularly cruise under the Arctic ice, their missiles can easily blanket the entire basin. The Kola Peninsula and the Murmansk region bristle with armaments; the United States maintains an important air base at Thule in northwest Greenland, another in Iceland and of course several in Alaska; the Canadians are beginning to supplement their passive defense systems in the north with more active components. How will all of this military activity affect scientific research in the region, much of which has military motivations and applications to begin with?

And what of the role of Sweden and Finland, cut off from direct frontage on the Arctic Ocean but projecting well north of the Arctic Circle and sharing many of the problems of the Arctic environment? Should they not have a voice in determining the future of the Arctic? This question will grow in importance as air and water pollution in the region become more significant. Because of the special characteristics of the Arctic environment, virtually everything related to the sea – resource exploitation, surface and subsurface navigation, environmental protection, criminal jurisdiction, to name but a few – requires special consideration and perhaps special legal arrangements. Small wonder that UNCLOS III left these matters to be handled by the circumpolar powers.

Antarctica and the Southern Ocean

If the north polar region is complicated, the polar politicogeographical situation in the southern hemisphere is much more so. For one thing, though the Arctic Ocean is a semi-enclosed sea with only a few relatively narrow passages to the south, the Southern Ocean surrounds the Antarctic continent but has no clearly definable northern limit. The most frequently suggested limit is the Antarctic Convergence (also called the Polar Frontal Zone), a zone about 50km (30 miles) wide between 55° and 62° South latitude where the cold Antarctic water sinks below the warmer waters of the Atlantic, Pacific and Indian Oceans. Some observers have suggested the parallel of 40° South, but it seems more reasonable to select a natural rather than a geometric boundary here. The one that may be most appropriate is the Subtropical Convergence (also called the Subantarctic Convergence), the southernmost reach of the warm, southward-flowing currents of the oceans to the north. It lies at approximately 40° South and displays more latitudinal stability than the Antarctic Convergence. Today, however, regardless of the merits or draw-backs of these and other suggested northern limits of the Antarctic region, the only one with any legal status is 60° South latitude.

The area enclosed by this line is the area covered by the Antarctic Treaty, under which a system of governance has been established that has no parallel or precedent anywhere in the world. The Antarctic Treaty was negotiated and signed in Washington in 1959 by the 12 countries that had been most active in Antarctic exploration and scientific research during the International Geophysical Year (1957–8). It went into effect in 1961 and is reviewable in 1991. As of mid-1989, an additional 27 countries had become parties to the treaty and 13 of them had joined the original 12 as Consultative Parties (CPs). These are the decision-making States, those that exercise whatever powers of governance exist in the region.

The treaty is short, only 14 articles. To summarize it briefly: Antarctica is to be used for peaceful purposes only; no military activities of any kind are permitted, though military personnel and equipment may be (and are) used

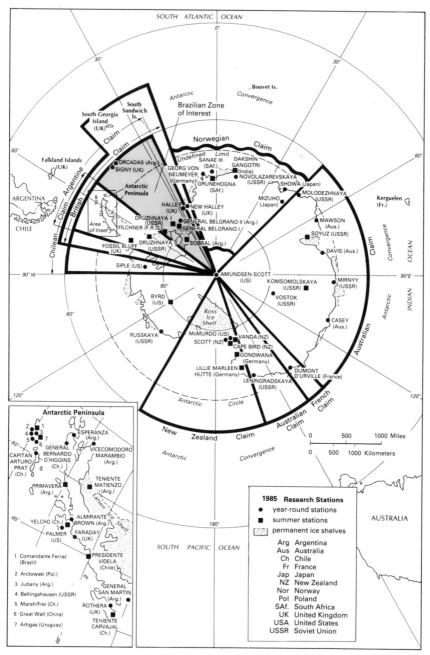

Figure 10.1 Antarctic research stations and territorial claims, 1989. The territorial claims shown on this map have remained unchanged for several decades. The formal Brazilian declaration of an Antarctic "zone of interest" is recent, but Brazilian geopoliticians, including the geographer Therazinha de Castro, have long been urging a Brazilian claim in the region. Research stations are constantly opening and closing; maps of Antarctic "settlements" must be updated much more frequently than comparable maps of other parts of the world. This is a characteristic of a rapidly changing frontier region, as Antarctica is.

for scientific purposes. Freedom of scientific investigation and cooperation shall continue. Scientific program plans, personnel, observations, and results shall be freely exchanged. No prior territorial claim is recognized, disputed, or established and no new claims may be made while the treaty is in force. Nuclear explosions and disposal of radioactive waste are prohibited. All land and ice shelves south of 60° South are covered, but not the high seas of the area. Observers from treaty States have free access to any area and may inspect all stations, installations, and equipment. Treaty States shall meet periodically to exchange information and take measures to further treaty objectives, including the preservation and conservation of "living resources." These consultative meetings shall be open to contracting parties that conduct substantial scientific research in the area. Disputes are to be settled peacefully, ultimately, if necessary, by the International Court of Justice.

Although all terms of the treaty are being faithfully carried out (with the possible exception of Article 9.1(f) concerning "preservation and conservation of living resources in Antarctica"), some matters the treaty omits are now emerging as problems. Among them is the validity of the treaty system itself as the instrument of governance in Antarctica. Some developing countries, led by Malaysia, are attempting to apply the common heritage principle to Antarctica and thereby supplant the treaty system with UN or some other international jurisdiction. This effort is being stoutly resisted by the treaty powers, though they have modified the system somewhat to accommodate the criticisms of Malaysia and others as described on p. 104.

Another problem is the question of sovereignty. Seven countries have claimed wedges of Antarctica, and three of the claims overlap. Most are based at least in part on the sector theory, but this is not an accepted principle of international law and Norway objects to it so adamantly that it has refused to set northern or southern limits to its claim. As in the Arctic, the status of ice in the Antarctic is unclear. And Article 4.2 states, "No new claim, or enlargement of an existing claim, to territorial sovereignty in Antarctica shall be asserted while the treaty is in force." Does this mean that no claimant may claim any maritime zones off its land claim? If such zones are permitted, where do they begin and end? Do the EEZ claims of Argentina and Chile in the Southern Ocean have any validity?

Still other questions are relevant to the political geography of the sea. The treaty intentionally makes no mention of economic activities on land or at sea within the treaty area, though all high seas rights are protected. In addition, while the treaty does include among its "principles and objectives" the "preservation and conservation of living resources in Antarctica," it says nothing about other kinds of environmental degradation such as air pollution or the accumulation of junk on the land, ice or water. Nor does it provide any environmental standards or any means of enforcing such standards should they be adopted by the Consultative Parties. Finally, although the treaty specifically says in Article 1 paragraph 1, "Antarctica shall be used for peaceful

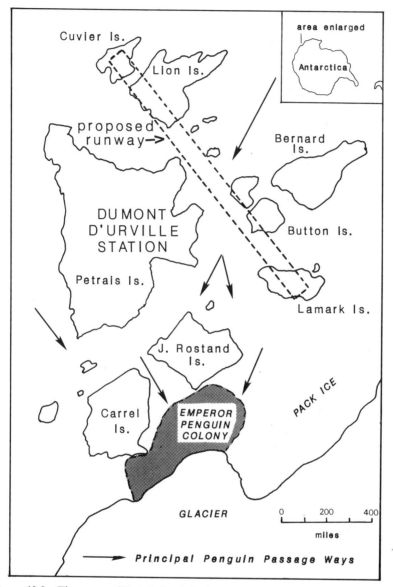

Cuvier Is.

Lion Is.

area enlarged

Antarctica

proposed
runway→

Bernard
Is.

DUMONT
D'URVILLE
STATION

Petrais Is.

Button Is.

Lamark Is.

J. Rostand
Is.

Carrel
Is.

EMPEROR
PENGUIN
COLONY

PACK ICE

GLACIER

0   200   400

miles

→ Principal Penguin Passage Ways

Figure 10.2   The proposed French airstrip in Antarctica. All of the islands on this map are covered by colonies of sea birds and the emperor penguin colony shown here is one of the few colonies — and the most accessible — of the least common of penguin species. The penguin population here has declined by more than half since the 1950s, the decline being especially rapid since the French in January 1983 began blasting, bulldozing and helicoptering to build an airstrip that many observers consider quite unnecessary. The French recently suspended construction under heavy pressure from environmental groups. (Antarctic and Southern Ocean Coalition 1985, frontispiece)

purposes only" and prohibits military bases, maneuvers, weapons testing and other such activities in the treaty area, there have been disquieting reports of remote sensing devices and satellite tracking facilities functioning within the treaty area. It seems unlikely that nuclear submarines on patrol or engaged in maneuvers have invariably remained north of 60° South, and the Anglo-Argentine war over the Falkland Islands in 1982 generated other questions about military activities in the treaty area.

All of these matters bear directly on the relationship, if any, between the Antarctic Treaty and the United Nations Convention on the Law of the Sea. This issue was studiously avoided at UNCLOS III and is still a long way from resolution. Leaving aside questions of sovereignty, do the Consultative Parties have the legal right to adopt rules binding on non-parties to the treaty? If so, how can these rules be enforced? If not, who does have the power to adopt and enforce such rules? How much of the Law of the Sea is applicable south of 60° South? Specifically, what about commercial activities and protection of the marine environment in the Southern Ocean? Harvesting of seals, whales, finfish and krill has been carried out in the Southern Ocean for more than a century, until recently with little or no regulation except for the last few decades by the International Whaling Commission. Already the search for oil and gas on the Antarctic continental shelf has begun. Vessels and aircraft bearing tourists – not scientists or support personnel or official observers, but ordinary tourists – have been sailing and overflying the Southern Ocean and reaching the continent itself, sometimes getting into serious trouble, disturbing the pristine environment and even being involved in fatal accidents.

The Consultative Parties of the Antarctic Treaty System have been mindful of all of these questions and were active participants in UNCLOS III. Three of them, however, the USA, UK and the Federal Republic of Germany, refused to sign the 1982 Convention and may not become parties to it for a long time, if ever. Other parties to the Antarctic Treaty have taken other positions vis-à-vis the Convention. They have remained stoutly opposed, however, to any internationalization of Antarctica, any application of the common heritage principle to the treaty area. Meanwhile, they have taken action through the treaty mechanisms to deal with specific problems that have arisen in connection with activities in Antarctica and the Southern Ocean. Thus they have adopted Agreed Measures for the Conservation of Antarctic Fauna and Flora (1964), the Convention on the Conservation of Antarctic Seals (1972) and the Convention on the Conservation of Antarctic Marine Living Resources (1980), and in 1988 a treaty establishing a minerals regime applicable to both the continent and its shelf.

It would appear, therefore, that without any design or plan, the Antarctic Treaty powers are in the process of developing a kind of regional economic zone/continental shelf in the Southern ocean, constructed piecemeal in response to perceived needs and tailored to the special conditions of the Antarctic environment. It seems likely that this process will continue. Under

Table 10.1 Members of the Antarctic Treaty System (1989)

Original consultative parties (12)	Date of Ratification or Accession
Claimant states (7) (The date in parentheses is the date of the formal claim.)	
United Kingdom (1908)	May 31 1960
Norway (1939)	Aug 24 1960
France (1924)	Sep 16 1960
New Zealand (1923)	Nov 1 1960
Argentina (1943)	Jun 23 1961
Australia (1933)	Jun 23 1961
Chile (1940)	Jun 23 1961
Non-claimant states (5)	
South Africa	Jun 21 1960
Belgium	Jul 26 1960
Japan	Aug 4 1960
United States	Aug 18 1960
Soviet Union	Nov 2 1960
Later consultative parties (13) (The date in parentheses is the date of becoming a consultative party.)	
Poland (Jul 29 1977)	Jun 8 1961
German Dem. Rep. (Oct 5 1987)	Nov 19 1974
Brazil (Sep 12 1983)	May 16 1975
Fed. Rep. Germany (Mar 3 1981)	Feb 5 1979
Uruguay (Oct 7 1985)	Jan 11 1980
Italy (Oct 5 1987)	Mar 18 1981
Peru (Oct 1987)	Apr 10 1981
Spain (Sep 21 1988)	Mar 31 1982
PR China (Oct 7 1985)	Jun 8 1983
India (Sep 12 1983)	Aug 19 1983
Sweden (Sep 21 1988)	Apr 24 1984
Finland (Oct 1989)	May 15 1984
Rep. of Korea (Oct 1989)	Nov 28 1986

Non-consultative (14)	Date of Accession
Czechoslovakia	Jun 14 1962
Denmark	May 20 1965
Netherlands	Mar 30 1967
Romania	Sep 15 1971
Bulgaria	Sep 11 1978
Papua New Guinea	Mar 16 1981
Hungary	Jan 20 1984
Cuba	Aug 16 1984
Greece	Jan 8 1987
Korea Dem. Rep.	Jan 21 1987
Austria	Aug 25 1987
Ecuador	Sep 15 1987
Canada	May 4 1988
Colombia	Jan 31 1989

Figure 10.3 Beardmore South Field Camp, Antarctica. In October 1982 Malaysia proposed to the UN General Assembly that the United Nations focus its attention on Antarctica. This began an attempt to apply the common heritage principle, developed for seabed resources, to Antarctica and to supplement or replace the Antarctic Treaty System (ATS) with United Nations or other broad international control of the region. As one means of defusing this potentially explosive situation, the United States National Science Foundation organized a workshop in Antarctica for a representative group, including diplomats and UN officials. The photograph shows some of the workshop participants in January 1984. It is perhaps coincidental but noteworthy nevertheless that since the workshop the ATS has tried in a number of ways to accommodate some of the demands of Malaysia and others; ten additional States have acceded to the Antarctic Treaty, half of them developing countries, and nine countries have become Consultative Parties, including China and three developing countries; and the annual debates on the subject in the General Assembly have dwindled considerably. (Mitchell Werner)

the impact of the United Nations General Assembly debates on Antarctica and of the constant nagging of nongovernmental environmental organizations, the CPs are opening up the system to more observers from nonparties (including nongovernmental organizations), are distributing their documents more widely, are making it easier for States to become CPs and are moving toward establishing some kind of formal, if tenuous, link with the UN.

The Antarctic Treaty system has been remarkably successful. The continent is the only large land area on earth that has never been militarized. The treaty region has been kept free of nuclear weapons, tests and wastes. Scientific research has been unfettered by artificial restraints and has been most effective. Cooperation in research and in the administration of the treaty has been exemplary. Despite some lapses and some weaknesses, protection of the

Antarctic environment has generally been good. There have been, despite conflicting territorial claims and a sizeable unclaimed area, no territorial conflicts as have been seen in virtually every other part of the globe. Antarctica is no longer remote, however, no longer isolated from political, technological, demographic, environmental and economic changes sweeping the world. The world is intruding forcefully into the last frontier on earth.

The peoples of the world will be ill-served if a system that has worked so well for so long is torn apart by fratricidal struggles within or by constant attack from without. It has already demonstrated some flexibility. More will be necessary in order to reach an accommodation with the Law of the Sea, with the movement for a new international economic order and with the United Nations. All this must be done, for, in the words of the preamble of the Antarctic Treaty, "it is in the interest of all mankind that Antarctica shall continue forever to be used exclusively for peaceful purposes and shall not become the scene or object of international discord."

11 *Military uses of the sea*

The sea has been a battlefield for millennia and there is no reason to believe that it will cease to be so in the foreseeable future. Yet this major use of the sea is excluded from special consideration in the United Nations Convention on the Law of the Sea and was not even an agenda item at UNCLOS III. There are a number of reasons for this. Firstly, one of the basic premises of UNCLOS III was that it was to produce a treaty that would, among other things, "promote the peaceful uses of the seas and oceans," in the words of the preamble of the Convention, and the Conference was difficult and prolonged enough without bringing military issues into it. Second, military uses of the sea have historically been covered by the law of war, not the Law of the Sea, and the rules of naval warfare are spelled out in great detail and well understood by navies around the world. Third, the USA and the USSR, which today deploy most of the world's naval strength, were in substantial agreement on the matter and did not want it discussed in such a large and diffuse forum over which they could exercise little control.

Nevertheless, military, or security, issues were not absent at UNCLOS III. In fact, they permeated and at times dominated the deliberations, and permeate the 1982 Convention. Many delegations included naval officers and some of these officers played prominent roles in the Conference and in the public discussions during and after it. Even though no section of the United Nations Convention on the Law of the Sea is devoted to it, then, the Law of the Sea has a military aspect that deserves some attention here. Since we are geographers, rather than political scientists or strategists and since we are political, not military, geographers, we will confine the discussion to selected themes that fall within the ambit of political geography.

The needs of navies and of coastal States

The navies of the superpowers, and to varying degrees, the navies of other countries, have today five principal missions:

1 *Strategic deterrence.* They strive to project such power (or the image of

power) that potential enemies will be deterred from attacking the countries that employ them.

2 *Projection of power ashore.* They carry military strength far from home and project it onto the shores of enemies, by bombarding coastal targets, launching aircraft that attack targets on shore and landing troops and supplies.

3 *Control of the uses of the sea.* They keep open sea lanes they use and protect their own shipping, fishing fleets and other marine activities, and try to interdict or destroy the comparable activities of an enemy.

4 *Naval diplomacy.* They "show the flag" in foreign waters, engage in highly visible maneuvers and otherwise serve as an instrument of foreign policy by encouraging friends and intimidating adversaries.

5 *Surveillance and intelligence-gathering.* They cruise just outside a country's territorial waters and use sensitive devices to spy on inland, coastal and maritime communications and other activities, explore the bottom of the sea, track submarines, etc.

All of these activities are characterized by mobility and flexibility, two qualities that make navies so valuable, even in an age of intercontinental ballistic missiles. It was the primary function of all of the naval officers at UNCLOS III – and many of the civilian diplomats – to preserve for their navies as much mobility and flexibility as possible so that they could continue to carry out their multiple missions in the interest of their national security. Thus the countries with significant blue-water navies generally strove to maintain the narrowest possible bands of national jurisdiction offshore, to retain freedom of navigation through and over straits (or "choke points"), to exempt warships from pollution controls and other limitations on their construction and operations, and to make a clear distinction between military activities and scientific research in the EEZs.

The coastal States have a different perspective. They look outward and see threats to their national territory and to their maritime operations from the fleets of hostile or potentially hostile States. Thus, they tend to require conditions quite the reverse of countries with powerful, wide-ranging navies. They tend to feel threatened by the very weapons that supposedly provide "strategic deterrence" to the outbreak of war; they certainly do not want anyone else's power projected onto their shores; they would prefer to retain control of their own uses of the sea; they object to naval diplomacy as much as to any other form of intimidation; and they do not want anyone spying on them.

We therefore have a classic confrontation of conflicting interests: many countries with little power wanting to restrain naval activity and a few countries with massive power wanting maximum freedom for their naval activity. However, in international relations, as in so many other areas of life, things are rarely as simple as they appear to be. In this case, the positions of the negotiating States were seldom either emphatic or constant. The naval powers had other interests as well: security of their own coasts, distant-water fishing fleets, marine scientific research, petroleum companies operating off the coasts of other countries, and so on.

Some coastal States also had other concerns: their own aspirations to become naval powers, the costs of enforcing restrictive rules in extensive jurisdictional waters, alliances with naval powers, and others. In fact, some delegations to UNCLOS III, notably that of the USA, were riven by disputes among contending interests and themselves resembled mini-conferences. Add to this the changing circumstances outside the Conference and it is not hard to understand that even the most adamant positions were subject to erosion and that negotiation and compromise were possible. This is reflected in the United Nations Convention on the Law of the Sea.

On the whole, military interests in the sea are well protected, in some ways even more so than before 1973. Warships, for example, have liberal rights of passage through the territorial sea, straits and archipelagic sea lanes; protection fron interference by coastal States exercising their continental shelf rights; exemption from all environmental protection provisions; and exclusion of military activities and marine scientific research from regulation of activity on the deep seabed. In the EEZ, there are no prohibitions on military maneuvers, testing of weapons or emplacement of devices of a military nature except, of course, where such activities can be construed as constituting a threat to the peace, good order or security of the coastal State, activities which are prohibited by the UN Charter and by the LOSC itself.

States are specifically permitted to lay pipelines and cables within any EEZ and on any continental shelf, in accordance with both customary and conventional international law, even if such pipelines and cables have actual or potential military value. States are prohibited from emplacing artificial islands, installations and structures on other States' continental shelves, but the word "devices" was deliberately excluded from this prohibition. The emplacement of military devices on or under the shelf is therefore permitted, so long as it does not violate any other obligation under international law.

Military aircraft retain preexisting rights of navigation even though their courses may take them over an EEZ, a strait or an archipelagic sea lane. And a State may, if it wishes, exclude from the mandated dispute settlement procedures "disputes concerning military activities, including military activities by government vessels and aircraft engaged in non-commercial service ..." (Art. 298(b)).

Peaceful uses of the sea

The Convention articles, as well as the preamble, are sprinkled with references to peaceful uses of the sea. For example:

- The high seas shall be reserved for peaceful purposes. (Art. 88).
- The Area shall be open to use exclusively for peaceful purposes by all States ... (Art. 141).
- Marine scientific research in the Area shall be carried out exclusively for peaceful purposes ... (Art. 143).
- [M]arine scientific research shall be conducted exclusively for peaceful purposes. (Art. 240).
- In exercising their rights and performing their duties under this Convention, States Parties shall refrain from any threat or use of force against the territorial integrity or political independence of any State.... (Art. 301).

Does this mean that military activities are prohibited on, under and/or over the sea? Not at all. In the absence of outright hostilities, there is a tacit understanding that all military activities at sea are undertaken by a State in exercise of its sovereign right of self-defense. Except where specifically prohibited (e.g. in territorial waters, straits and archipelagic sea lanes), naval maneuvers and other activities are permitted so long as they do not involve a "threat or use of force" against another country.

Clearly this raises questions of interpretation. When, for example, does "gunboat diplomacy" become a threat of force? When does exercising a warship's legal right to traverse another State's territorial waters (or even EEZ) become a deliberate and dangerous provocation? These are legal and political matters beyond the scope of this book, but they do illustrate the impossibility of clearly delineating peaceful and non-peaceful uses of the sea. More geographic is the series of attempts in recent decades to mark off portions of the earth's surface – both land and sea – as "nuclear-free zones" and "zones of peace."

The first proposal for a nuclear-free zone came in 1957, when Adam Rapacki, foreign minister of Poland, proposed a zone in Central Europe, on both sides of "the Iron Curtain," that would be freed and kept free of all nuclear weapons. Nothing came of the Rapacki Plan, but similar proposals followed. The first actual nuclear-free zone was established in Latin America by the 1967 Treaty of Tlatelolco. Since then, other such zones have been proposed for the Balkans, the Adriatic and the Mediterranean; Africa; Northern Europe; the Middle East and South Asia, among others. None of these has actually been legally established by treaty, however. Four other areas have, so far, been declared nuclear-free zones to date: Antarctica (1959), outer space (1967), the seabed (1970) and the South Pacific (1985–6).

Even more sweeping, and more idealistic, are the proposals that have been made over the years for zones of peace. The first of any consequence in modern times was the Declaration of South-East Asia as a Zone of Peace, Freedom and Neutrality, adopted by the Association of Southeast Asian Nations (ASEAN) in November 1971 – in the midst of the Vietnam War. In the following month the United Nations General Assembly adopted the Declaration of the Indian Ocean as a Zone of Peace, and in November 1986 it adopted the Declaration of a Zone of Peace and Co-operation of the South Atlantic.

None of these zones of peace has actually been implemented yet, and it is unclear what practical effect they will, or could, have on naval activities, nuclear or conventional, or how they can actually promote peace in the world. Perhaps, combined with the exhortations in the 1982 Law of the Sea Convention and the United Nations Charter itself, they will have some deterrent effect on powers that might be tempted to disturb the peace of these areas. As for the nuclear-free zones, the first three have apparently been scrupulously observed. The most recent, in the South Pacific, has been rejected by the United States, which has reserved the right to sail nuclear-powered and nuclear-armed vessels through the region, and by France, which insists on the same rights and in addition continues to test nuclear weapons in French Polynesia.

Military zones in the sea

Although States have had difficulty establishing zones of peace in the sea, they seem to have had little trouble establishing and implementing military zones of various kinds. States do, in fact, regularly control, restrict, interdict and prohibit the exercise of freedom of navigation on and over the high seas and the EEZ. One of the ways they do so is by establishing zones within which they exercise full or partial military control and within which passage and even the entry of vessels and aircraft is limited in some way. Some appear to be permanent, or at least destined to persist until the world is at last at peace. Others are *ad hoc*, created to meet particular needs and then dissolved when the need has passed. What follows is only a suggestive list with brief descriptions that can serve to illustrate the point.

Permanent
Air defense identification zones, maintained by the United States, Canada, the Soviet Union and other countries within which aircraft must identify themselves or risk being shot down. Since these zones extend far beyond territorial waters, they could be considered as interference with the freedom of overflight guaranteed under both the EEZ and the high seas regimes.
The North Korean 50-mile defense zone which, for all practical purposes, is a territorial sea and clearly violative of the widely accepted 12-mile limit. Unlike other countries that claim territorial seas of more than 12 miles in breadth, North Korea freely uses military force to deter entry by unauthorized vessels.

Ad hoc
Military test zones, such as the Pacific missile ranges announced from time to

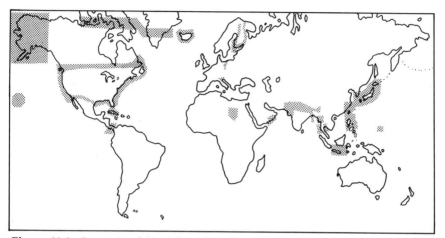

Figure 11.1 International air defense identification zones, 1985. Note the extensive marine areas covered by these zones, not only adjacent to the superpowers, but to many other States as well. (Compiled from maps in US Department of Defense Flight Information Publications AP/1, AP/2 and AP/3)

time by the Soviet Union and the United States and the area around Muroroa Atoll in the South Pacific, used by France for testing nuclear weapons. Since these areas are cleared of shipping by military vessels and aircraft before weapons tests, they may be considered violations of high seas and EEZ freedom of navigation.

Exclusion zones, such as those established by the UK in 1982 during its successful attempt to retake the Falkland Islands from Argentina. First the UK declared a maritime exclusion zone of 200 nautical miles around the islands within which Argentine vessels and aircraft could proceed only at their peril. This soon became a total exclusion zone which applied also to ships and aircraft of countries supporting Argentina. After the cessation of hostilities, this was replaced by a 150-mile "protection zone" which Argentine ships and aircraft may not enter without permission. During the Iran–Iraq War of the 1980s, Iran declared and attempted to enforce a similar exclusion zone in the Persian Gulf.

The Beira Patrol, operated by the UK from 1966 to 1975 off the coast of then-Portugese Mozambique under authorization of the United Nations Security Council to enforce the Council's economic sanctions against Rhodesia. Specifically, the task of the patrol was to prevent tankers from delivering petroleum to Beira, whence it could be piped to Rhodesia. It was not notably successful.

The United States "quarantine" of Cuba, undertaken in 1962 during the Cuban Missile Crisis to prevent offensive missiles from reaching Cuba by sea. This was a unilateral move, without authorization by the Security Council but under the dubious cloak of approval by the Organization of American States.

It is much too early to evaluate the ultimate impact of the new Law of the Sea on military uses of the sea, or of military needs and activities on the interpretation and implementation of the Convention. It seems likely at this point that military activities on, under and over the sea will be affected only slightly. It must be borne in mind, however, that it is by no means certain that all questions about such matters as transit passage and maneuvers within the EEZ will always be decided in favor of the naval forces of the superpowers, regardless of their unrivalled firepower. Nor can we predict the effects of the growing navies of middle-rank powers, or the development of new ocean shipping routes, or of new technology, or of artificial satellites in space or of many other factors likely to have an impact on the military uses of the sea. We can only hope that those countries that have navies use them constructively, to keep the peace at sea rather than to disrupt it.

12 Regional arrangements in marine affairs

The United Nations Convention on the Law of the Sea includes the words "region", "subregion", "regional" or "subregional" or some combination of these words in no fewer than 21 articles. Nowhere in the Convention, however, are these words defined. Furthermore, all references to regional or subregional cooperation or other action are permissive or hortatory in nature rather than mandatory. Thus, while the concept of "regional" activity and "region" as an arena of activity is paid due deference in the Convention, its imprecision must raise questions about its meaning and its long-term impact upon the Law of the Sea, man's uses of the sea and international relations in general.

What is a region?

The region as a geographic concept evolved in the late 19th century and has since gone through many phases. There is no generally accepted succinct definition of the term. This is not for lack of effort, however. Combining the most useful elements of a number of attempted definitions, we can explain the concept in this manner: A region is a portion of the earth's surface characterized by some kind of homogeneity in its core. The homogeneity is most intense at the center and fades toward ill-defined boundaries or even into zones of transition into other regions. A region cannot exist in isolation but only in relation to the total structure of which it is a component, and even within the region there may be a number of interrelated features which constitute the homogeneity which defines the region. Any number of subregions may be identified within each region.

What kind of homogeneity can define a region? Any kind at all that can be mapped. Until very recently "region" applied only to land areas, sometimes with local water bodies included; the concept of a marine region is relatively new, and will undoubtedly evolve in a variety of unforeseeable ways. Since a region is defined by the observer for a particular purpose, we may expect to see marine regions of many kinds developing in coming decades.

If we consider, as the delegates to UNCLOS III apparently did, that "regional cooperation" in marine affairs is good, then we must try to determine the nature and extent of a region in which such cooperation should take place. This is not easy. For one thing, the land-based concepts of region are difficult, if not impossible, to transfer to the sea. There are few recognized historical, economic, cultural or physical features in the sea to guide us in identifying and delimiting regions. Second, merely extending political boundaries from the land into the sea, not an easy task in itself, would probably in most cases create political areas that lack the cohesion and homogeneity necessary for the designation " region". In fact, they are likely, in most cases, to do violence to any rational concepts of ecological unity and administrative practicality. Furthermore, verticality is much more pronounced here than on land, and we may have to devise vertical as well as horizontal regions. The outer limits of the continental shelf and the exclusive economic zone, for example, will rarely coincide, and there may be a need for different regimes governing the harvesting of pelagic and demersal fish. There are other differences between marine and land regions, but the point is clear: Establishing rational and useful marine regions will be difficult.

Lewis Alexander has identified three principal types of marine regions: 1) *physical regions*, "differentiated from other areas on the basis of coastal configuration;" 2) *management regions*, "representing situations where there is a well defined management problem, capable of being handled as a discrete issue;" and 3) *operational regions*, "sites of one or more regional arrangements."[1]

It is quite evident that these types of regions are often very difficult to reconcile, even if they can be clearly identified. It is very likely, for example, that certain physical features of the sea currently ignored by nearly everyone but marine scientists will become prominent as management regions come to be established for various purposes. Major currents, the Subtropical Convergence, and extrusions of polymetallic sulfides come immediately to mind as features that may require special attention in the future. Ideally, perhaps, operational regions should coincide with management regions; that is, a regional organization should have complete jurisdiction over a complete marine feature, as, for example, pollution control or fishery conservation. But there are many kinds of pollution and many kinds of fish, and it is difficult at this point to see how any regional marine organization could deal effectively with all of them without taking on a supranational character which, given the intensity of nationalism prevalent in the world today, is unlikely to be acceptable for a long time to come.

Marine regions in the Law of the Sea Convention

The Convention seems to be ambivalent about the role of true internationalism. On the one hand, it strikes out boldly toward supranationalism by establishing an International Seabed Authority and an International Tribunal for the Law of the Sea, both of whose decisions are binding upon States parties. On the other hand, throughout the Convention, even in most provisions concerning the Authority and the Tribunal, the emphasis is clearly on the rights, privileges, benefits and duties of States. Regions and regional organizations are mentioned only in relation to certain situations and functions that are or should be transnational. Thus regionalism seems to be treated here as a purely pragmatic instrument for achieving the goals of both individual States and the global society without any implication of a greater role for it in the future. The number and prominence of such mentions in the treaty text, however, are indications of the importance attached to the region and to cooperation among States.

In the Convention there are seven references to regional arrangements or agreements, six to regional organizations, three to regional rules or standards, six to regional programs or centers and thirteen occurrences of "region" or "regional" standing alone and apparently referring to some undefined portion of the sea. With regard to regional arrangements, agreements, organizations, rules, standards, programs and centers, there is no problem of definition. Each is or will be defined in terms of its participants. If a number of countries (usually but not necessarily contiguous) sign an agreement, or create an organization to deal with a particular marine matter, then the portion of the sea falling under the jurisdiction of such agreement or organization would constitute a region. Similarly with rules, standards, programs and centers; the portion of the sea covered by each would constitute a region. The problem here is not one of definition, but one of a multiplicity of regions of many sizes and characteristics – horizontal, vertical and functional – overlapping and perhaps competing with one another in a dizzying array of jurisdictional disputes. How far we have come from Elizabeth Mann Borgese's description of the sea as "an indivisible ecological whole!"

Added to this is the problem of defining "region" when it stands alone in the Convention, referring either to an area or to a level of international organization. Presumably each of these "regions" will be defined *ad hoc* by whichever States are most directly involved. But this raises numerous questions whose answers may be quite elusive. To pose only a few that come readily to mind:

1   Must a State with more than one seacoast share equally in the benefits and burdens of managing or administering all portions of the sea contiguous to its coasts? Consider, for example, France, Panama, Egypt and Malaysia. Each fronts on two seas or oceans, and perhaps many more marine regions.

2   May a State with no seacoast at all derive benefits from the seas of *all* its coastal neighbors, even if they front on different seas or oceans? Bolivia, Switzerland, Uganda and Nepal are only a few obvious examples.

3   Since the sea and the land are inextricably intertwined in the coastal zone along most shorelines, does a marine "region" include the coastal zone as well? If so, how far inland? What would be the relationship between the jurisdictions of the State and the region? Is an *ad hoc* agreement on this matter the best solution from the standpoint of the marine environment, or economic efficiency, or the welfare of residents of the coast?

4   What criteria would be used to define "region" in each of the thirteen cases in which the term is unmodified in the Convention? *Ad hoc* criteria again? Suppose some States in what would appear to be a logical "region" according to reasonable criteria see things very differently or refuse to participate in a particular regional activity: Is the area covered by agreement among the others still a region? Does it fall under the relevant provisions of the Convention?

In order to glimpse the kinds of problems that will very likely arise in the interpretation and application of the concept of a marine region, it might be useful to make a cursory survey of some existing regional activities.

A survey of some existing marine regions

Until quite recently the largest classes of both agreements and organizations concerned with regional marine affairs have been those devoted to the conservation of the "living resources" of the sea. Since most of these marine animals live within 200 nautical miles of the coast, they will come under national jurisdiction as more and more countries establish 200-mile exclusive fishing or economic zones as sanctioned by the Convention. Already a number of the regional conservation organizations have become moribund or have been dissolved, while others are in the process of redefining their missions and functions.[2]

The second largest group of agreements and organizations are those concerned with preventing and controlling pollution of the sea. Those still functioning date from 1969 or later. A large number were created between 1971 and 1976; more have been added since then and undoubtedly more will be formed in coming years. Whereas there is one United Nations agency primarily concerned with fisheries – the Food and Agriculture Organization (FAO) – there are two with jurisdiction over pollution – the International Maritime Organization (IMO) and the United Nations Environment Programme (UNEP). The former deals primarily with ship-generated pollution of the sea, while the latter concentrates on land-based pollution. Of all the numerous agreements, organizations and other anti-pollution activities, the most relevant to our theme is the UNEP Oceans and Coastal Areas Programme. (Until late 1986, this was known as the Regional Seas Programme.)

Under this program, action plans have thus far been adopted for eleven areas of the global sea: the Mediterranean, Persian Gulf, Caribbean, West and Central Africa, East Asian Seas (including the Straits of Malacca), Red Sea and Gulf of Aden, South Asian Seas, Southeast Pacific, South Pacific, East Africa and Southwest Atlantic. How many of these can reasonably be classified as "regions" in accordance with the criteria spelled out at the beginning of this chapter is questionable. In designating its "regional seas," UNEP had to consider a number of factors. Among these, the basic one was vulnerability to environmental dangers. If this were the only factor, clearly the Baltic Sea should have been UNEP's first priority, since by the early 1970s it was already considered by many ecologists to be a "dying" sea. Other considerations, however, must enter into decisions of this kind, especially administrative feasibility and political acceptability.

It would seem at first glance that semi-enclosed seas constitute "natural" marine regions, and Article 123 may be interpreted as an expression of that assumption. But it may also be noted that Article 123 specifies only three subjects for "co-ordination" among the littoral States: The living resources of the sea, protection and preservation of the marine environment, and scientific research. If even these limited objectives could be achieved in any semi-enclosed sea, then it might justifiably be considered as a "region" in a general sense. Short of that degree of cooperation among the States surrounding a semi-enclosed sea, however, it is difficult to see how any such area of the sea can properly be considered a region, despite its logic based on configurations on a map. How close even the semi-enclosed seas come at present to being real regions, to say nothing of such vague areas of the global sea as "The Southwest Atlantic," may be judged from the following survey of various areas of the sea.

South and Southeast Asia

The dominant intergovernmental organization in this region is the Association of Southeast Asian Nations (ASEAN). Founded in 1967 by Indonesia, Malaysia, the Philippines, Singapore and Thailand, it did not really become active until 1976 and did not expand its membership until 1984, when it was joined by newly independent Brunei Darussalam. Notably absent from its membership are Myanma (formerly Burma) and the three States in what was French Indochina: Kampuchea, Laos and Vietnam. Notably absent also, from the list of its political, economic, social and technical activities, is any major concern with the sea, although it does conduct some marine-related activities.

Considering that this historic region is one of the culture hearths of human civilization, faces both the Indian and Pacific Oceans, and has great maritime traditions and heavy dependence on the sea today, this seems strange. There is some bilateral cooperation in marine affairs, between Thailand and Malaysia and between Malaysia and Brunei, but compared with their cooperation in land-based activities, it is not consequential and there is still no subregional commitment to cooperation in marine and maritime activities. This situation is tacitly recognized by UNEP, which, with the countries involved, has developed action plans for "South Asian Seas" and "East Asian Seas" but not for Southeast Asian Seas.

The situation is similar in South Asia. The world's newest subregional organization of general competence is the South Asian Association for Regional Cooperation (SAARC). It was formally launched in August 1983 on the separate initiatives of Bangladesh and Nepal, and its other members are Bhutan, India, the Maldives, Pakistan and Sri Lanka. Its secretariat is located in Kathmandu, Nepal. Its list of concerns – meteorology, sports, science, transport, health, rural development, culture, etc. – do not include any marine or maritime activities. Perhaps before long, either or both of these organizations will turn attention to the sea, but meanwhile there are two intergovernmental programs in the area that are concerned exclusively with such activities.

The First Conference on Economic, Scientific and Technical Co-operation in Marine Affairs in the Indian Ocean in the Context of the New Ocean Regime (IOMAC-I) was held in Colombo, Sri Lanka, in the summer of 1985. Since then, IOMAC has come to mean Indian Ocean Marine Affairs Cooperation. It is not an elaborate organization but it does have a secretariat in Colombo to coordinate its activities. Participation (not membership) is open to all Indian Ocean littoral and hinterland States (including the land-locked ones), as well as other countries active in the Indian Ocean. Competent international organizations and NGOs also participate. To date, participation has included 46 countries, among them Bhutan, China, the German Democratic Republic, Greece, Malawi, Nepal, Romania, Uganda, UK, USA and Zimbabwe. Twenty-eight UN organs, other intergovernmental organizations and NGOs have also been represented at IOMAC activities. These activities fall under eight headings: Marine science and ocean services, marine technology and training, living resources, non-living resources, maritime transport, communications and management (which includes the Law of the Sea and information services). This is an ambitious program. If successful and sustained, it could bring enormous benefits to the peoples in, on and near the Indian Ocean. If not, it would join the lengthy list of deceased Asian intergovernmental organizations and would probably be replaced before long by something similar but stronger.

A more specialized effort is the Bay of Bengal Programme for Fisheries Development (BOBP). It covers seven littoral States: Bangladesh, India, Indonesia, Malaysia, the Maldives, Sri Lanka and Thailand. (Note again the absence of Myanma [Burma].) Its main project is described as "small-scale fisherfolk communities in the Bay of Bengal." It is executed by the FAO and funded by Denmark and Sweden. It helps the artisanal fishermen by developing techniques, technologies and methodologies through pilot activities. It began in 1987 and is scheduled to last five years. It succeeds an earlier BOBP project, development of small-scale fisheries, which terminated in 1986. The current project includes safety at sea, mariculture, sail design and many other prosaic but pragmatic and very important matters.

The South Pacific

Since the South Pacific is mostly water, it is reasonable to assume that any regional organization would be at least marginally concerned with the sea. In fact, this is true. The region has three intergovernmental organizations: The South Pacific Commission (SPC), the South Pacific Forum (Forum) and the Forum's subsidiary, the South Pacific Bureau for Economic Co-operation (SPEC). Their functions and membership overlap considerably, and in 1987 they were considering merger. Meanwhile, they cooperate quite well.

The SPC is essentially an international technical assistance agency with an advisory and consultative role, but it does not have any political functions or bias and does not operate large aid programs or common services. It has 27 members of equal status, consisting of 22 island countries and five metropolitan States, with a modest secretariat staff in Nouméa, New Caledonia. Among its larger accomplishments is the 1986 Convention for the Protection of the Natural Resources and the Environment of the South Pacific Region.

The Forum is very different. It is the annual meeting of the heads of government of the independent and self-governing countries of the South Pacific, currently totalling 13, with the Federated States of Micronesia as an observer. Besides SPEC, which was established in 1973 with headquarters in Suva, Fiji, and which is heavily engaged in facilitating cooperation and consultation among the members in fisheries, transporation and many other matters, the Forum has had two major successes so far. In 1979 it established the Forum Fisheries Agency (FFA) to collect and disseminate to its members the whole range of information relevant to fisheries in the South Pacific, and it assists its members in negotiating fisheries treaties. This latter activity culminated in the adoption in April 1987 of the Treaty on Fisheries Between the Governments of Certain Pacific Island States and the Government of the United States, which resolved some very difficult problems relating to activities of US tuna fishermen in the region. Its other major achievement was its strong influence in the negotiation and adoption in August 1985 of the South Pacific Nuclear-free Zone Treaty which entered into force on 11 December 1986.

Finally, a newer organization should be noted briefly. The Co-ordinating Committee for Off-Shore Exploration in the South Pacific (CCOP/SOPAC) promotes and undertakes pre-exploration studies of the non-living marine resources of the region. It receives support from a number of intergovernmental agencies such as the United Nations Development Programme, the Commonwealth and the European Community, and from a number of individual countries. It is too early to evaluate its effectiveness, but its very effort to investigate the commercial potentials of such things as nodules, cobalt crusts, polymetallic sulfides, placer minerals, marine phosphates, black coral and wave energy is an indication of a commitment to reduce the region's reliance on fishing as a foreign exchange earner.

The Caribbean

Like the South Pacific, the Caribbean consists mostly of many small islands scattered over an area of sea that have a long history of association and a short history of cooperation. The first intergovernmental organization in each, the Caribbean Commission and the SPC, was initiated by the colonial powers. But the South Pacific learned from the mistakes of the Caribbean and did not repeat them. The Caribbean has seen many organizations come and go, from the Anglo-American Caribbean Commission (1942–6) through the Federation of the West Indies (1958–62) and the Caribbean Free Trade Association (CARIFTA, 1968–73) to the present Caribbean Community (CARICOM, 1973–), and Organization of Eastern Caribbean States (OECS, 1981–). Each had its successes and failures. The present organizations seem to be more durable than the earlier ones, but there has still not evolved any region-wide organization that includes not only the Commonwealth Caribbean, but also the Spanish-, French- and Dutch-speaking islands as well as those under US rule. Indeed, there seem to be more factors and forces dividing these islands than uniting them, to say nothing of the "Wider Caribbean," which includes the Gulf of Mexico and all of the mainland countries fronting on both bodies of water. There has traditionally been more conflict than cooperation among them.

Now it is possible that – at last – the sea might become a unifying factor. Because the Caribbean is so much more compact than the South Pacific, was exposed to industrialization and international trade much earlier, is more densely populated, more strategically located and under greater pressure from outside forces, there are many areas in which cooperation is not only indicated but required. These would include maritime boundaries and zones, pollution and conservation, navigation, fisheries, marine scientific research and transfer of marine technology, and security.

There does seem to be, at least in the Commonwealth Caribbean, a recognition of the need for joint marine policies and for common marine services. Funds and technical assistance are coming from outside to support moves in this direction. In 1983 UNEP was the midwife of the Convention for the Protection and Development of the Marine Environment of the Wider Caribbean Region and its protocol on oil spills. UNEP's Caribbean Action Plan consists of 66 environmental management projects, progressing at varying rates of speed. And there is an incipient Eastern Caribbean Regional Coast Guard Service, with the participation of Antigua, Barbados, St Lucia and St Vincent. But this is only a beginning. There is still very far to go to effective regional management of the marine environment and human activities therein.

Western Europe

Finally, to the grouping of States that one would assume to be closest to developing true marine regionalism: The European Community. We introduced its Common Fisheries Policy in Chapter 9; here it receives more attention.

It appears that in some ways, as the EC has expanded both numerically and functionally, its supranational character has diminished. Agreements seem harder to reach. Its growing pains appear to be more severe and lasting longer than many observers had expected. In oil and gas production, shipping, pollution control and environmental policy, agreement is still far off. At UNCLOS III the EC seldom spoke on any issue of importance with a strong and united voice until near the end. And at the very end, two of its strongest members – the UK and West Germany – refused to sign the Convention and have still not become parties to it. Yet they did not vote against it, they did sign the Final Act of the Conference and they are observers at the deliberations of Prepcom. With Hamburg designated as the site of the Law of the Sea Tribunal and other advantages to be derived from full participation in the Convention, it seems likely that West Germany will accede to it in due course and the UK will probably do so as well. It would then become easier for all members of the EC to reach agreement on important ocean-related issues. Meanwhile, for all its gaps and weaknesses, the Common Fisheries Policy is a signal achievement.

The Policy first took shape in 1970, and evolved continually, slowly before 1977 when the members extended their fishing limits to 200 miles, more rapidly since. There is by now a considerable body of fisheries law designed to implement the Policy and clarify its ambiguities. In recent years the EC as a whole has been the third largest fishing power in the world, after Japan and the USSR, landing about 10 percent of the total world catch. But there are many differences among the fishing industries of the members. For example, notwithstanding the imposition of 200-mile exclusive fisheries and economic zones in waters traditionally fished by them, France, West Germany, the UK, Portugal and especially Spain, still field collectively a sizable distant-water fleet. On the other hand, the middle-water and inshore fleets have grown and become more active off the shores of Greece and Ireland and parts of Denmark, France, Portugal, Spain and the UK. It has thus been difficult to determine which types of vessels should have preferential access to the Community fisheries, and which should receive what kind of Community help in conversion, modernization, laying-up or scrapping of fishing boats no longer needed. In addition, the member countries tend to prefer different species of fish: The British prefer whitefish, such as cod, haddock and whiting; West Germany concentrates on redfish, France and Spain on tuna, and Portugal and Spain on sardines. This leads to different interests in management, marketing and trade.

Finally, the members tend to fish in different parts of the world, and Spain and Portugal, alone among the members, frequently do so through joint ventures. This also leads to management problems. Pricing has been another problem because of fluctuating supply and demand of various species, and the importation of cheap fish from Canada, Iceland, Norway and the Faeroe Islands. Small wonder that the EC had such great difficulty in developing a Common Fisheries Policy and grand wonder that they did it!

There is still much to be done and many obstacles to overcome before the EC can develop common policies and practices in other marine affairs, and there is no assurance that they will be successful. Some disquieting signs are the withdrawal of Greenland from the Community on 1 January 1985 in order to gain control over its own fisheries, the very limited success of environmental cooperation in the North Sea, and the failure of the Community to agree on a uniform breadth for the territorial sea. Perhaps, as in the Indian Ocean, a very cautious optimism is appropriate for the future of marine regionalism in Western Europe.

The future of marine regionalism

The four situations surveyed above represent the most successful attempts at establishing marine regions so far and, as we have seen, none can yet be considered especially complete or effective. We could add some smaller groupings, but again the outcome would not be very encouraging. The Gulf Cooperation Council (officially the Cooperation Council for the Arab States of the Gulf) for example, consists of Bahrain, Kuwait, Oman, Qatar, Saudi Arabia and the United Arab Emirates. While officially dedicated to a program of cooperation in various fields, they have not ranked any marine affairs very highly among them and in truth their chief concern has been a common defence against external threats, most immediately from Iran. Then there is the Comisión Permanente del Pacífico Sur (Permanent Commission of the South Pacific). It was formed in 1952 by Chile, Ecuador and Peru primarily to lobby for a 200-mile EEZ. Once that goal was accomplished, it turned its attention exclusively to marine scientific research and environmental concerns. It was thereupon, in 1979, joined by Colombia. It is very active and does some good work, but there is no sign that it plans to expand its activities into other marine affairs or become the nucleus of a new marine region.

Marine regionalism is still weak around the world. Even in semi-enclosed seas there is still more competition than cooperation, even less joint marine activity and virtually no integration. The reason for this seems plain enough. Despite the numerous references in the LOSC to marine regions and despite the explicit evaluation of such regions as "good," States are not rushing eagerly to strengthen old regional arrangements or to formulate new ones. To the contrary, some existing arrangements seem to be disintegrating. One observer analyzes the situation this way:[3]

Much of the discussion of marine regionalism suggests that nations which do not eagerly pursue regional arrangements in environmental protection or fishing or other sectors somehow fail to see their own true interests ... Such discussions fail to acknowledge the very real value of national control over national affairs.

Because of the bias in favor of maintaining national control, nations generally shun regional co-operation unless it can be demonstrated that very substantial benefits would likely result from such co-operative efforts. For advocates of co-operation, then, one major task is to undertake serious analysis of where co-operation is most likely to be fruitful, and to show just how fruitful it is likely to be.... Another avenue is to show that the costs and risks of co-operation can be limited.

While recognizing all of the objections and impediments to marine regionalism, however, it is still possible to anticipate its growth and development in the decades to come. The slow development and primitive nature of marine regionalism, evident from the foregoing survey of some contemporary regions, indicate only that the goal is difficult to achieve, *not* that it is undesirable. Nor are the obstacles insuperable. Although the LOSC does not mandate or even define marine regions, it does recognize that they are more than desirable; they are essential. They are essential in part because of conditions created by the Convention itself. They are essential because many marine situations are simply too big to handle on a national basis and yet for a variety of reasons are not amenable to global approaches. Among all the possible alternatives, that of the marine region seems most appropriate, most likely to bring the greatest benefits and fewest risks and costs, in at least a large number of aspects of marine affairs, most especially those for which regional approaches are urged in the Convention.

13 The future political geography of the sea

No one can foresee what the political geography of the sea will be like a generation from now. Certainly it will be very different. Very soon, in order to convey realistic images of national size and relationships, political maps of the world and even of individual coastal and island States will have to show zones of maritime jurisdiction as well as land territory and internal waters. People in the 21st century will surely be much more sea-minded than we are today, and the sea will play a greater role in international relations than ever before in history. Control over marine resources, over important shipping lanes, over the best sites and means for generating energy from the sea, will most certainly contribute substantially to a radical realignment of power relationships among States.

Although we cannot know what the future political geography of the sea will be like, we can identify many factors that will influence it. We have already considered some of them, but several deserve additional attention. Again, the topics selected for extended discussion are by no means the only ones deserving of it, nor are these topics discussed in the detail warranted by their importance. Here we can only suggest, and hope that the reader will probe further into this fascinating, unknown world.

Science and technology

We have already seen how the scientific analysis of polymetallic nodules in the late 1950s led directly to Arvid Pardo's proposal for an International Seabed Authority, and his common heritage principle. But this was only the beginning. The sea contains immense deposits of many other kinds of minerals whose potential value is incalculable. Phosphorus, for example, is found abundantly in the form of phosphorite nodules, phosphatic pellets and sands, phosphatic muds and consolidated phosphate beds. Already some of these deposits are being tested – near New Zealand and Baja California – and before long, as the need for agricultural fertilizers to help feed a hungry planet increases, the temptation to mine them will increase accordingly. The same is true of the hot brines and metalliferous muds so abundant in the Red Sea and perhaps elsewhere. Sudan and Saudi Arabia are cooperating in the research and development of this potential resource that may soon be strengthening their economies.

Other subsurface mineral deposits are likely to have even greater impact on national economies in the intermediate future. Sands and gravels, already so important in building and construction, may well be exploited further to construct artificial islands for a variety of uses, as we discussed in Chapter 9. Radiolarian oozes on the seabed, especially in the Pacific, may someday provide new ceramic materials for many industrial uses. Polymetallic sulfides may provide relatively cheap iron, zinc, copper, manganese, gold, silver, aluminum, platinum and other metals from hot water vents on the seabed. So far, the only large deposits found lie along the East Pacific Rise and the Galapagos Rise, off the west coast of North America from British Columbia to California, and near the Galapagos Islands of Ecuador, but more are likely to be found near Japan and elsewhere. Manganese crusts, hot springs on seamounts and other mineral deposits may also become important resources.

Besides minerals, the sea can yield virtually unlimited energy. Small-scale tidal power plants have been in use for centuries, and the French have been operating a larger one on the estuary of the Rance River since the mid-1960s. The Soviets operate a small one near Murmansk. Others have been proposed for the Bay of Fundy and for the estuaries of the Mersey and Severn Rivers in England. The economic, political and environmental considerations of tidal power have so far prevented its full development, and it is possible that the benefits will never outweigh the costs. But even if tidal power never becomes important, other forms of energy from the sea might.

The most publicized in recent years, and perhaps the one with the greatest potential, is ocean thermal energy conversion (OTEC). It is based on the extraction of heat from an exchange between the warm surface waters of tropical and subtropical seas (typically 20°C–25°C), and the cold Antarctic waters underlying them at depths of about 1000 meters (a constant 4°C). The USA experimented with OTEC plants in Hawaii and Guam in the early 1980s and considered developing a very much larger one for the Gulf of Mexico. Japan, India, the Netherlands, Sweden and France have all begun OTEC projects and the UK may soon join in the research and development. Although the potential for OTEC is far larger than that for tidal power, there remain a variety of economic, political, environmental and technical problems to be solved before its potential can be realized.

Still other forms of ocean-generated energy have been investigated: wave power, ocean currents, salinity gradients, density gradients, biomass energy, hydrogen production at sea, ocean wind energy and mining undersea coal deposits by gasification. It is likely that none of these processes will ever produce a significant proportion of the world's energy; it is even possible that all of them together will never be truly important worldwide. But there could be major political implications if only a few countries are relieved of their dependence on imported conventional fuels by the development of reasonably efficient and economic marine energy.

Science and technology are likely to influence the world politicogeographical and geopolitical situations in other ways, for example food production. We have already mentioned the possibility of developing new fisheries and utilizing non-traditional species of marine animals. There are other possibilities. Science and technology can help us to locate and harvest fish and other "living resources" more efficiently, with reduced losses of the harvest *en route* to market, reduced undesirable incidental catches of dolphins and other unwanted species, and more complete utilization of the animals harvested. Then there is the potential for greater utilization of corals, sponges and most important, seaweeds. Although we have been consuming and using many kinds of seaweeds for many centuries, there are species scarcely or never used and new uses for the traditional crops. Mariculture is also likely to grow dramatically in coming decades, not just small-scale farming of shellfish and finfish in nearshore waters, but larger-scale ranching and cultivation of marine animals even far out to sea, in their natural habitats. Again, a shift of food production areas can be important politically.

Seawater itself is already yielding a variety of minerals, but many more can be extracted, even with existing technology, when market conditions are satisfactory. Desalinization of seawater is currently practiced widely around the world, but at high cost. Efforts have been underway in Israel, the USA and elsewhere to develop a cheap, practical and reliable method of doing this on a large scale. How this could change the barren shores of Northwest Africa or Western Australia or Southern South America! And we need only mention the extraction of hydrogen from seawater (even now an easy process) to provide unlimited fuel for the production of fusion energy; imagination can carry on from there.

There is no doubt at all that artificial islands (discussed in Chapter 9) will become important, perhaps commonplace, within a few decades; so will improved underwater communications, such as fiber-optic cables and exceedingly accurate marine surveying and measuring devices, undersea archaeology using new submersibles and tracked vehicles, commercial submarine transport and more sophisticated space-sea communications going far beyond remote sensing.

None of this is going to happen quickly, which is a very good thing. We still have time to think through the potential implications of every new scientific discovery and technological development. Every incremental change will somehow change existing relations among States and we would be well advised to keep alert to onrushing science and technology or risk entering the 21st century still thinking in 19th-century terms.

Questions of territory and limits

In Chapters 3 through 5 we discussed at length both the customary and the new methods of determining a State's jurisdiction or sovereignty over the sea, as well as the newly designated International Seabed Area. The discussion, however, did no more than lay the groundwork for an understanding of how profoundly the world will be changed by the extension of national jurisdiction far out to sea. Now we return to the theme for a few suggestions about the kinds of changes likely to result.

First, there is no reason to believe that the United Nations Convention on the Law of the Sea is the last word on the question of national jurisdiction in the sea. In truth, there seems to be little to prevent a State from either extending its jurisdiction far beyond the arbitrary 200-mile limit, or gradually converting its economic zone into a territorial sea, or both. In fact, during the Law of the Sea Conference the term "creeping jurisdiction" became widely used to describe these possible activities, and the term is now common in the professional literature. And, as anyone familiar with the modern history of the Law of the Sea could have predicted, it is the CEP countries (Chile, Ecuador and Peru) that continue to lead the way.

On September 12, 1985 Chile claimed a 350-mile continental shelf around Easter and Sala y Gomez Islands. A week later, as noted in Chapter 3, Ecuador claimed the entire Carnegie Ridge between its claimed 200-mile territorial seas off its coast and around the Galapagos Islands as part of its continental shelf, thus giving it uninterrupted, exclusive right to all resources in a zone extending some 700 nautical miles from the coast of its continental territory. These are only two of what will undoubtedly be numerous claims to jurisdiction to the outermost limits allowed by the UN Convention on the Law of the Sea (although it has not yet entered into force) and then, on the flimsiest pretexts, far beyond those limits. At the same time, jurisdiction within the EEZ going beyond that specified in the Convention is already being claimed by some States, even as far as authority to restrict navigation on various creative grounds. Unless something acts to restrain this aggressive nationalism, creeping jurisdiction will break into a gallop until the "world lake concept" becomes a reality and the high seas vanish altogether. This was a proposal made late in the 1960s to divide the entire global sea among coastal and island States along median lines between them. How *that* would change the political geography of the world!

Even without the problem of creeping jurisdiction, however, the delimitation of maritime boundaries under the new Law of the Sea is complex and important. Since Pardo's celebrated speech of 1967, the World Court alone – as we noted in Chapter 3 – has dealt with maritime boundaries in the North Sea (1969), the North Atlantic (1974), the Aegean Sea (1978), Libya and Tunisia (1982), the Gulf of Maine (1984), and Libya and Malta (1985), with still more such cases on its docket. Other major cases have been settled by arbitration: the UK and France (1977), the Beagle Channel (1984), Guinea and Guinea-Bissau (1985). There still remain scores of maritime boundaries to be delimited and the process of doing so may take decades. While in many cases there are no particular difficulties, there is no apparent urgency for countries with many problems and few resources to enter into complex negotiations over boundaries of little immediate importance to them. In other cases, however, there are active disputes.

It may be recalled that there are still unresolved boundary and territorial problems on land. Where such problems involve seacoasts, clearly settlement of maritime boundaries must await settlement of the terrestrial boundaries. Besides numerous disputed islands, there are: Venezuela's claim to two-thirds of Guyana, Guatemala's claim to all of Belize and the boundary dispute between El Salvador and Honduras. China claims much of the Soviet Far East, many of the boundaries of the United Arab Emirates are undefined, and the status of the Gaza Strip is uncertain, while Iraq and Iran fought an eight-year war begun in 1980 at least in part over their mutual boundary in the Shatt-al-Arab. The status of Western Sahara is unclear, while Namibia undergoes a transformation to independence from South Africa. Are Ciskei and Transkei sovereign States requiring maritime boundaries with South Africa, or still parts of the republic?

The USA alone still has about a score of unsettled maritime boundaries, including three with Canada (the Beaufort Sea, the Dixon Entrance and the Strait of Juan de Fuca) and two with the USSR (in the Chukchi Sea and the Bering Sea). It is the disputed islands that constitute the largest category of maritime boundary problems. They are found in all ocean basins. Even where the ownership of the islands themselves is not in dispute, there is frequently disagreement over proper baselines, the nature and extent of the continental shelf, whether the bits of land qualify as islands in the first place, the alignment of the boundary when islands are close to a coast of another State, and so on. Besides islands that protrude above the sea, numerous undersea features such as ridges, trenches, banks and reefs complicate the process of boundary-making.

Even if two opposite or adjacent States agree on the principles to be used in defining their maritime boundaries, there is still the question of whether there should be separate boundaries between their continental shelves and their EEZs, or whether one boundary should serve for both. So far, only between Australia and Papua New Guinea and between Australia and Indonesia are there separate boundaries for the shelf and the EEZs, because of special local circumstances. The trend seems to be toward single boundaries cutting vertically through air, sea, shelf and subsoil, as in the Gulf of Maine and Guinea-Guinea Bissau cases. In many cases, however, only the shelf or EEZ boundaries have been negotiated but not both, and we may still see more cases of a State having jurisdiction under or over another State's jurisdiction.

There are so many questions with regard to the actual drawing of straight baselines, determination of normal baselines, delimitation of the outer limit of

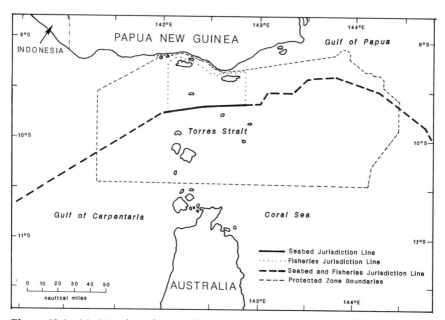

Figure 13.1 Maritime boundaries in the Torres Strait. This deviation from the general pattern of single maritime boundaries was made in order to protect the ancestral fishing rights of the Torres Strait islanders, who are culturally and ethnically distinct Australian nationals. This rather unusual concession to the rights of a small minority group in a sensitive region may bode well for future maritime boundary deliberations. (Annex 7 to the Torres Strait treaty between Australia and Papua New Guinea, 1978)

the continental shelf, determination of archipelagic waters and so on, that it is difficult to make useful generalizations about them now. Many of them will have to be settled *ad hoc*, at least until enough settlements have been reached for some general patterns of State practice and third-party decisions to have emerged to guide the parties in the remaining disputes.

International organization for the sea

In the UN Convention on the Law of the Sea there are dozens of articles in which reference is made to "appropriate" or "competent" international organization(s), and still others that include references to subregional or regional organizations. These articles are scattered throughout the Convention and its annexes and resolutions. They are both a device to integrate the constructive work of existing intergovernmental organizations into the new Law of the Sea, and an encouragement to States to work through either existing or new organizations to achieve the objectives of the Convention. Indeed, it is fair to say that even the most ardently nationalistic delegations at UNCLOS III recognized that there are marine and maritime matters that are

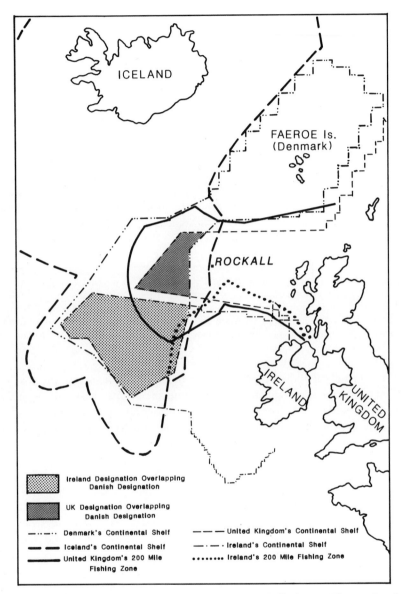

Figure 13.2 Disputed maritime boundaries on the Rockall Plateau. The overlapping claims of Denmark, Iceland, Ireland and the United Kingdom present a complex but by no means unique picture. Similar problems exist in the South China Sea, the East China Sea and the Bay of Bengal. They are much more difficult to resolve than the Gulf of Maine and Beagle Channel disputes discussed and illustrated in Chapter 4, and more such complex situations are likely to develop within the next decade. (Symmons 1986, pp. 2, 4, 5)

simply beyond the management capacity of even the richest and most powerful countries. Cooperation among States, between States and inter-governmental organizations and among these organizations is absolutely essential if utter chaos at sea is to be avoided.

We have already considered the International Seabed Authority and the International Tribunal for the Law of the Sea, which are to be established under the terms of the Convention, and the Preparatory Commission charged with developing their rules and procedures. It is probable that other inter-governmental organizations will be established in future to perform other functions. The experience of the existing "competent" and "appropriate" organizations deserves some attention at this point.

Generally, they can be placed into two categories: Those that are part of the United Nations system and those that are not. Among the former that are active in marine affairs are the UN itself, which has a large body within the Secretariat headed by an Under-Secretary General devoted to these matters, and various organs and specialized agencies such as the IMO. In 1983, IMO established in Malmö, Sweden the World Maritime University to provide advanced training for specialized maritime personnel, particularly from developing countries, in maritime safety and administration, protection of the marine environment and international shipping.

The International Labor Organization has a Joint Maritime Commission that deals with wages and working conditions of seamen. The Intergovern-mental Oceanographic Commission of UNESCO and the International Hydro-graphic Organization do basic research and survey work and coordinate similar work of member States. The Food and Agriculture Organization has a large and active Fisheries Division, and the Commission for Marine Meteor-ology of the World Meteorological Organization is concerned with the economic impacts of weather at sea, among other things. The World Health Organization and the International Atomic Energy Agency are concerned with marine pollution and disposal of radioactive wastes at sea. Together, the UN, FAO, UNESCO, WHO, WMO, IMO and IAEA sponsor the Joint Group of Experts on the Scientific Aspects of Marine Pollution (GESAMP).

The United Nations Conference on Trade and Development has a Ship-ping Division that, among other activities designed to help developing countries utilize the sea for shipping, has developed a Code of Conduct for Liner Conferences. The UN's regional commissions for Europe, Western Asia, Latin America and the Caribbean, Africa, and Asia and the Pacific sponsor various marine-related activities depending on the needs of their member States. And, of course, there is UNEP, discussed in Chapter 12, and the International Court of Justice, referred to in the preceding section.

Among the non-UN organizations, some are particularly noteworthy. The International Whaling Commission and the many regional and species-specific fisheries commissions are responsible for research on, and in some cases regulation of use of, their respective subjects. The International Maritime

Satellite Organization (INMARSAT) operates space satellites that provide basic information about the sea and its resources. The Commission for the Conservation of Antarctic Marine Living Resources was created by the parties of the Antarctic Treaty System for the purpose indicated by its title.

The purpose of this incomplete catalogue of intergovernmental organizations active in sea-related matters is to emphasize the point made in Chapter 12: The sea is so vast and complex that no State, no matter how large or rich or powerful, can benefit fully from it without cooperating with other States. Science and technology are constantly opening new vistas and raising new questions; maritime boundary problems cannot always be resolved bilaterally; pollution knows no national boundaries; transportation, communications, trade, conservation and the maintenance of peace and security all require international cooperation. In the future States will have to work together even more than they are doing now. This cooperation is both a result of and another catalyst for the changing political geography of the sea.

Non-governmental organizations

In his major address to the final session of the Third United Nations Conference on the Law of the Sea in Montego Bay, Jamaica, in December 1982, the Conference President, Ambassador Tommy T. B. Koh of Singapore, listed "those features of the negotiating process of this Conference which were productive ... " Included in his list of nine features was the following:

> I should also acknowledge the role played by the non-governmental organizations, such as the Neptune Group. They provided the Conference with three valuable services. They brought independent experts to meet with delegations, thus enabling us to have an independent source of information on technical issues. They assisted representatives from developing countries to narrow the technical gap between them and their counterparts from developed countries. They also provided us with opportunities to meet, away from the Conference, in a more relaxed atmosphere, to discuss some of the most difficult issues confronted by the Conference.

This remarkable and historic tribute to a group of private citizens was richly deserved. They came from several countries and represented several private, non-profit citizens' groups, both secular and of various religious denominations, to pool their talents and their energies at every session of the Conference to help the delegates prepare a fair and practical constitution for the sea. They produced a newspaper called *Neptune*, which is how the group got its name. This newspaper provided highlights and sidelights on the Conference and factual information for the delegates. It was demonstrably influential in shaping the wording of a number of provisions of the Convention, besides performing the educational functions and the good offices described by Ambassador Koh.

The Neptune Group broke up when UNCLOS III ended, and the members went back to their countries, organizations, and schools, usually to work on other major international issues – disarmament, Antarctica, ecology. A direct lineal descendent of the Neptune Group is the Council on Ocean Law in Washington, D.C. It continues the work of educating both the US Government and the general public about marine affairs. It produces a valuable newsletter, reproduces articles on marine affairs from other sources, and conducts seminars and conferences on marine issues. It is an excellent source of current facts and analysis in layman's language.

Other non-governmental organizations concerned with the sea are also doing fine work in research, education and, in some cases, lobbying legislators and government officials on sea-related issues. Some of the major environmental groups do this as part of their overall activity; the Sierra Club International, Greenpeace and the International Institute for Environment and Development are among the most prominent. Others specialize in marine affairs, e.g., the Oceanic Society and the Cetacean Society International. The Antarctic and Southern Ocean Coalition is made up of 160 NGOs in 30 countries and has headquarters in Australia and Washington, D.C.

The Law of the Sea Institute at the University of Hawaii and the International Ocean Institute of Malta both sponsor major annual conferences of specialists in marine affairs in various locales around the world and publish their proceedings. They also conduct workshops or seminars on specialized topics. The LOSI also publishes occasional papers, while the IOI conducts courses and training programs for government personnel and others working in or preparing for ocean-related positions.

The Scientific Committee for Oceanographic Research (SCOR) has some 22 active working groups, committees and panels studying and coordinating marine scientific research around the world. The International Union for the Conservation of Nature and Natural Resources (IUCN), based in Switzerland, has in recent years done more and more work in marine ecology.

These are only some of the numerous groups of private citizens working constantly in many countries to help shape the relations between law and politics on the one hand, and the sea and our uses of it on the other. They provide two-way channels of communication, transmitting information about the sea and marine affairs to their members and the general public, and the concerns and opinions of ordinary people to governments and intergovernmental organizations. Some of them send observers to meetings of UN bodies, the International Whaling Commission and the Antarctic Treaty Consultative Powers, among others. Any individual can join most of them and contribute actively to development of a better world, as did the women and men of the Neptune Group at UNCLOS III. One person can make a difference.

14 *We are the stewards of Neptune's Domain*

Roll on, thou deep and dark-blue Ocean – roll!
Ten thousand fleets sweep over thee in vain;
Man marks the earth with ruin – his control
Stops with the shore.
Lord Byron, Childe Harold IV

Alas, man is still marking the earth with ruin, but his control no longer stops
with the shore. One of the justifications advanced by proponents of a 200-mile
EEZ was that the high seas beyond a narrow belt of territorial waters had been
polluted and overfished because it was treated as common property, and
therefore subjected to "the tragedy of the commons." Therefore, the solution
to the problem was to let individual States undertake the management of the
most valuable parts of the sea. The fallacy in that argument was clear from the
beginning: If, in fact, the sea had been *res communis*, the property of all, then
every State, every seaman, every fisherman owned a share of it and should
have striven to protect that share by cooperating with all other owners in the
management of the entire commons. But the sea was *not* treated as a
commons; it was treated as *res nullius*, the property of no one, and thus open to
be raped and ravaged at will.

We are now entering the testing time. Will the new system work better
than the old non-system? A wise and experienced marine scientist expressed it
this way in a symposium on large marine ecosystems at the 1984 meetings of
the American Association for the Advancement of Science:

Now, what of the apparent contradictions of extended coastal jurisdiction on the one hand and the need for even greater international
cooperation in ocean research on the other? Two events of the past year,
El Niño and the possibility of a partial krill recruitment failure in the
Antarctic, illustrate the importance of the need for even greater international cooperation.... As new discoveries are made with greater
frequency now that research and assessment efforts are increasing within

the territorial seas, we will need to work even harder exchanging scientists, conducting joint research operations, and convening symposia and workshops to nurture international cooperation. I look forward to greater international participation in the open-ocean waters. We have an unprecedented opportunity for demonstrating to the world the great benefits in joint international studies of the oceans.[1]

More than international studies, we need international management of the sea based on those studies. As we have seen in this book, such management, on a regional and world-wide basis, is given mild encouragement in the United Nations Convention on the Law of the Sea. But one of the lessons to be learnt from the study of history and of political geography is that nationalism is probably the strongest force in the world, and is likely to remain so, at least for a while longer. Developing widespread and effective international cooperation in marine affairs will not be easy and it will not come soon. Advancing from cooperation to international management of the entire marine ecosystem, or even major portions of it, will be very much more difficult and take far longer. But it can be done.

Figure 14.1 Dr Arvid Pardo, who initiated the movement toward a new Law of the Sea with his dramatic proposal in 1967 that the resources of the deep seabed be considered the "Common Heritage of Mankind". He was an observer at the concluding session of UNCLOS III in Montego Bay, Jamaica, in December 1982, where this photograph was taken. In between, however, he watched sadly as his original idealistic proposal was transformed into something much less inspiring. We may hope that some day his dream will be realized. (Martin Glasssner)

One place to begin is by implementing promptly the kinds of cooperation in MSR and transfer of marine technology outlined in the Law of the Sea Convention. Another is to channel more resources to the regional and international organizations already doing fine work in marine affairs, including those described in Chapters 12 and 13. Another is to give international support to the establishment and maintenance of national marine parks, reserves and sanctuaries, perhaps through UNEP and UNESCO. All of these activities could be initiated immediately at little cost in cash or in national sensibilities.

The next step could well be the creation of a worldwide system of marine parks in international waters, managed by an international agency established for the purpose. The parks could include choice samples of all important submarine features and ecosystems, and of historical and cultural treasures. One could well be the entire Southern Ocean south of the Subtropical Convergence. Another could be in the central Indian Ocean, surrounding and including the Chagos Archipelago. Another in the Sargasso Sea. Others could preserve intact samples of spreading ridges, nodule fields, fracture zones, warm and cold currents, deep trenches and so on, just as various physical environments are preserved on land. The wreck of the *Titanic* would be a logical international historical sanctuary, and many more could be identified. If this is to be as useful a project as it could and should be, planning for it should begin right now, before it is too late, before man marks the sea with ruin.

A radical proposal? Not really. In fact, one can scarcely imagine a more conservative approach to our relation to the sea than recognizing that it is not ours to destroy, that it remains Neptune's domain and we are only its stewards. We must conserve it, restrain our instincts to exploit it until we understand it better, until we can administer multiple-use policies that will ensure continued productivity, continued absorptive capacity, continued enjoyment for all peoples. No one knows better than a political geographer with experience in international relations just how difficult it will be to take this course. For lying in wait all over the world, in, around and far from the sea, are radicals disguised as conservative lawyers, diplomats, government officials, corporation executives and bankers, radicals more interested in plunder than in conservation.

But these people, I believe, can learn the benefits to be gained from cooperation and conservation. Fortunately, others feel the same way. John Byrne expressed it this way:

> I am optimistic. I believe the future will be a period of a holding pattern on damage and a more enlightened approach to the conservation and management of both renewable and non-renewable resources.

As individuals, we can work alone, in groups and in non-governmental organizations to achieve this goal. And we can succeed.

Idealistic? Of course, but why not be idealistic? As that great British geographer Sir Halford J. Mackinder said in his 1919 book *Democratic Ideals and Reality*, "Idealists are the salt of the earth; without them to move us, society would soon stagnate and civilization fade."

Notes

Notes to Chapter 5

1 It was reported in December 1987, for example, that Japan plans to reinforce the island of Okinotorishima, which consists of two rocks only 5 meters in diameter, so as to preserve its claim to an EEZ and continental shelf around it.
2 See Alexander 1974. Dr Alexander is a Professor of Geography and Director of the Center for Ocean Management Studies at the University of Rhode Island. He also served for three years as the Geographer of the United States Department of State.

Notes to Chapter 7

1 This phrase was introduced by Elizabeth Mann Borgese, one of the outstanding leaders of the movement for a new Law of the Sea. Now at Dalhousie University in Halifax, Nova Scotia, she still organizes the annual Pacem in Maribus conferences, sponsored by the International Ocean Institute of Malta, and co-edits the *Ocean yearbook* published by the University of Chicago Press.
2 For more details, see Gold 1986. Professor Gold is an attorney and master mariner, a member of the Faculty of Law of Dalhousie University and Director of Dalhousie's International Institute for Transportation and Ocean Policy Studies.
3 For an excellent study of the current state of world fisheries, see FAO 1987a and comparable documents of subsequent years.

Note to Chapter 9

1 As pointed out in Chapter 1, the dispute settlement procedures of the Law of the Sea Convention are not particularly geographic and thus are not discussed in this book. They are important, however, and a number of sources in the selected bibliography discuss them.

Note to Chapter 10

1 Much of the material in these last two paragraphs is drawn from Butler 1988. Professor Butler teaches comparative law at the University of London and directs the Centre for the Study of Socialist Legal Systems, University College London.

Notes to Chapter 12

1 See Alexander 1978a.
2 The best recent discussion of this topic is FAO 1987b. It is quite comprehensive and contains much useful material in the form of summary charts.
3 Kent 1983, pp. 86, 90–1. George Kent is a Professor of Political Science and of Urban and Regional Planning at the University of Hawaii. His research focuses on the roles of ocean and food politics in development.

Note to Chapter 14

1   Byrne 1986, pp. 306–7. Dr Byrne is President of Oregon State University, where he was formerly Dean of the School of Oceanography. He has also served as US Commissioner to the International Whaling Commission. When he presented this paper he was serving as Administrator of the National Oceanographic and Atmospheric Administration (NOAA).

Selected bibliography

Akaha, T. 1985. *Japan in global ocean politics*. Honolulu: University of Hawaii Press and Law of the Sea Institute, University of Hawaii.

Alexander, L. M. 1963. *Offshore geography of northwestern Europe*. Chicago: Rand McNally.

Alexander, L. M. 1974. Regionalism and the Law of the Sea: the case of semi-enclosed seas. *Ocean Development and International Law* **2** (2), 151–86.

Alexander, L. M. 1978. *Regional co-operation in marine science*. Report prepared for the Intergovernmental Oceanographic Commission, the UN and the Food and Agriculture Organization.

Alexander, L. M. 1978a. Regionalism at sea: concept and reality. In *Regionalization of the Law of the Sea*, D. Johnston (ed.). Proceedings of the Eleventh Annual Conference of the Law of the Sea Institute, Cambridge, Mass.

Alexandersson, G. 1982. *The Baltic straits*. The Hague, Boston and London: Martinus Nijhoff.

Amin, S. H. 1986. *Marine pollution in international and Middle Eastern Law*. Glasgow: Royston.

Anand, R. P. 1976. *Legal regime of the seabed and the developing countries*. New Delhi: A. W. Sijthoff.

Anderson, L. G. 1977. *Economic impacts of extended fisheries jurisdiction*. Ann Arbor, Mich.: Ann Arbor Science.

Antarctic and Southern Ocean Coalition 1985. *Background paper on the French airfield at Pointe Géologie, Antarctica*. Sydney and Washington.

Archer, C. & D. Scrivener (eds). 1986. *Northern waters: security and resource issues*. Totowa, NJ: Barnes & Noble.

Attard D. J. 1987. *The exclusive economic zone in international law*. Oxford, England: Clarendon Press.

Barnaby, F. 1983. Military uses of the oceans. *Impact of Science on Society*. pp. 421–32.

Bartell, J. J. (ed.). 1982. *The Yankee mariner & sea power – America's challenge of ocean space*. Los Angeles: University of Southern California Press.

Baxter, R. R. 1964. *The law of international waterways, with particular regard to interoceanic canals*. Cambridge, Mass.: Harvard University Press.

Beck, P. J. 1987. *The international politics of Antarctica*. New York: St Martin's Press.

Bergesen, H. O. A. Moe & W. Østreng 1987. *Soviet oil and security interests in the Barents Sea*. New York: St. Martin's Press.

Birnie, P. 1985. *International regulation of whaling: from conservation of whaling to conservation of whales and regulation of whale-watching*. (2 vols.) London, Rome, New York: Oceana.

Blake, G. 1987. *Maritime boundaries and ocean resources*. London and Sydney: Croom Helm.

Bloomfield, L. P. 1981. The Arctic: last unmanaged frontier. *Foreign Affairs* **60** (1), fall, 87–105.

Boczek, B. A. 1984. Global and regional approaches to the protection and preservation of the marine environment. *Case Western Reserve Journal of International Law* **16** 39ff.

Bonner, W. N. & R. I. Lewis Smith. 1985. *Conservation areas in the Antarctic.* Cambridge, England: Scientific Committee on Antarctic Research.

Booth, K. 1985. *Law, force and diplomacy at sea.* London, Boston and Sydney: George Allen and Unwin.

Borgese, E. M. 1986. *The future of the oceans: A report to the Club of Rome.* Montreal: Harvard House.

Bouchez, L. J. 1964. *The regime of bays in international law.* Leyden, The Netherlands: A. W. Sythoff.

Bowett, D. W. 1979. *The legal regime of islands in international law.* Dobbs Ferry, NY: Oceana.

Broder, S. & J. Van Dyke. 1982. Ocean boundaries in the South Pacific. *University of Hawaii Law Review* **4** (1), 1–59.

Brooks, D. L. 1984. *America looks to the sea; ocean use and the national interest.* Boston and Woods Hole: Jones and Bartlett.

Brown, E. D. 1984–6. *Seabed energy and mineral resources and the Law of the Sea.* (3 vols.) London: Graham and Trotman.

Burmester, H. 1982. The Torres Strait Treaty: ocean boundary delimitation by agreement. *American Journal of International Law* **76** (2) 321–49.

Butler, W. E. 1978. *Northeast Arctic passage.* Alphen aan den Rijn, The Netherlands: Sijthoff and Noordhoff.

Butler, W. E. 1985. *The Law of the Sea and international shipping: Anglo-Soviet post-UNCLOS perspectives.* New York, London and Rome: Oceana.

Butler, W. E. 1988. The legal regime of the Soviet Arctic. In *The Soviet Maritime Arctic*, Woods Hole Oceanographic Institution, pp. 9–10. Proceedings of workshop held 10–12 May 1987 at Woods Hole, Mass.

Byrne, J. 1986. Large marine ecosystems and the future of ocean studies: a perspective. In *Variability and management of large marine ecosystems*, L. M. Alexander & K. Sherman (eds). Boulder, Colo: Westview Press.

Caflisch, L. C. 1978. Land-locked states and their access to and from the sea. *British Yearbook of International Law* **49**, 71–100.

Canada and the Sea 1980. Special issue of *Canadian Issues* **3** (1). A publication of the Association for Canadian Studies.

Carlisle, R. P. 1981. *Sovereignty for sale: The origins and evolution of the Panamanian and Liberian flags of convenience.* Annapolis, Md: Naval Institute Press.

Center for Ocean Management Studies. 1985. *Antarctic politics and marine resources: critical choices for the 1980s.* Kingston, RI: University of Rhode Island, Center for Ocean Management Studies.

Champ, M. A. & P. Kilho Park. 1982. *Global marine pollution bibliography.* New York: IFI/Plenum.

Charney, J. I. (ed.) 1982. *The new nationalism and the use of common spaces; issues in marine pollution and the exploitation of Antarctica.* Totowa, NJ: Allenheld, Osmun.

Cervenka, Z., (ed). 1973. *Land-locked countries of Africa.* Uppsala, Sweden: Scandinavian Institute of African Studies.

Chiu, H. 1985. Some problems concerning the application of the maritime boundary delimitation provisions of the 1982 United Nations conference on the Law of the Sea between adjacent or opposite states. *Maryland Journal of International Law and Trade* **9** (1), 1–17.

Churchill, R. R. 1987. *EEC fisheries law*. Dordrecht, Boston and Lancaster: Martinus Nijhoff.
Churchill, R. R. & A. V. Lowe. 1989. *The Law of the Sea*. 2nd edn. Manchester, England: Manchester University Press.
Constans, J. 1979. *Marine sources of energy*. New York: Pergamon Press.
Coull, J. R. 1972. *The fisheries of Europe; an economic geography*. London: G. Bell & Sons.
Couper, A. D. 1985. The marine boundaries of the United Kingdom and the Law of the Sea. *The Geographical Journal* **151**(2), 228–36.
Cusine, D. J. & J. P. Grant (eds). 1980. *The impact of marine pollution*. London: Croom Helm; Montclair, NJ: Allanheld, Osmun.
Cuyvers, L. 1984. *Ocean uses and their regulation*. New York: John Wiley and Sons.
Cuyvers, L. 1986. *The Strait of Dover*. Dordrecht, Boston and Lancaster: Martinus Nijhoff.

Dahmani, M. 1987. *The fisheries regime of the exclusive economic zone*. Dordrecht, Boston, Lancaster: Martinus Nijhoff.
Dyer, I. & Chryssastomidis C. (eds). 1984. *Arctic technology and policy*. Washington, DC: Hemisphere.

Earney, F. C. F. 1980. *Petroleum and hard minerals from the sea*. New York: John Wiley and Sons.
East, W. G. 1960. The geography of land-locked States. *Transactions and Papers* **28**, 1–22. Institute of British Geographers.
Eckert, R. D. 1979. *The enclosure of ocean resources: economics and the Law of the Sea*. Stanford, Calif.: Hoover Institution Press.
Edeson, W. R. & J. F. Pulvenis. 1983. *The legal regime of fisheries in the Caribbean region*. Berlin: Springer-Verlag.
El Baradei, M. 1982. The Egyptian-Israeli peace treaty and access to the Gulf of Aqaba: a new legal regime. *American Journal of International Law* **76** (3).
El-Hakim, A. A. 1979. *The Middle Eastern States and the Law of the Sea*. Syracuse, NY: Syracuse University Press.
Epting, J. T. 1980. National marine sanctuary program: balancing resource protection with multiple use. *Houston Law Review*. **18**, 1037–59.

FAO 1979. *Fisheries development in the eighties*. Rome: FAO.
FAO 1987a. *World fisheries situation and outlook*. COFI/87/2. (Committee on Fisheries.)
FAO 1987b. *Activities of International organizations concerned with fisheries*. Fisheries Circular. no. 807. Rome: FAO.
Feldman, M. B. 1983. The Tunisia-Libya continental shelf case: geographic justice or judicial compromise? *American Journal of International Law* **77** (2), 219–38.
Feldman, M. B. & D. Colson. 1981. The maritime boundaries of the United States. *American Journal of International Law* **75** (4), 729–63.
Fernie, J. 1977. The development of North Sea oil and gas resources. *Scottish Geographical Magazine* **93** (1), 21–31.
Fincham, C. & W. van Rensburg. 1980. *Bread upon the waters; the developing Law of the Sea*. Jerusalem, London: Turtledove.
Finn, D. P. 1982. *Managing the ocean resources of the United States: the role of the federal marine sanctuaries program*. Berlin, Heidelberg, New York: Springer-Verlag.
Francioni, F. & T. Scovazzi (eds). 1987. *International law for Antarctica*. Milan: Dott Giuffre Editore.
Friedheim, R. L. (ed.). 1979. *Managing ocean resources*. Boulder, Colo: Westview.
Friedmann, W. 1971. *The future of the oceans*. New York: George Braziller.

Fulton, T. W. 1976. *The sovereignty of the sea*. Millwood, New York: Kraus Reprint.

Glassner, M. I. 1970. *Access to the sea for developing land-locked States*. The Hague: Martinus Nijhoff.

Glassner, M. I. 1978. Transit rights for land-locked States and the special case of Nepal. *World Affairs* **140** (4), 304–14.

Glassner, M. I. 1985a. Regionalism and the new Law of the Sea. In *Essays on the new Law of the Sea*, B. Vukas (ed.), pp . 409–28. Zagreb: Institute of International Law and International Relations, University of Zagreb.

Glassner, M. I. 1985b. The view from the near north: South Americans view Antarctica and the Southern Ocean geopolitically. *Political Geography Quarterly* **4** (4), 329–42.

Glassner, M. I. (ed.) 1986a. The new political geography of the sea. Special issue of *Political Geography Quarterly* **5** (1).

Glassner M. I. 1986b. *Bibliography on land-locked States*, Dordrecht, Boston, Lancaster: Martinus Nijhoff.

Gold, E. 1981. *Maritime transport – the evolution of international marine policy and shipping law*. Toronto: D. C. Heath.

Gold, E. 1988. The control of marine pollution from ships: responsibilities and rights. In *The Law of the Sea: what lies ahead?* T. A. Clingan, Jr. (ed.), 276–96. Proceedings of the 20th Annual Conference of the Law of the Sea Institute, 21–24 July 1986, Miami, Florida. Honolulu: Law of the Sea Institute.

Gopalakrishnan, C. 1984. *The emerging marine economy of the Pacific*. Boston, Mass.: Butterworths.

Govindaraj, V. C. 1974. Land-locked States – their right to the resources of the seabed and the ocean floor. *Indian Journal of International Law* **14** (3–4) 409–24.

Graham, R. & D. W. Huff. 1983. *Annotated bibliography of underwater and marine park related initiatives in northern latitudes*. Waterloo, Ontario: University of Waterloo, Department of Recreation.

Griffiths, F. 1987. *Politics of the Northwest Passage*. Kingston and Montreal: McGill-Queen's University Press.

Gullion, E. A. 1968. *Uses of the seas*. Englewood Cliffs, NJ: Prentice-Hall.

Halberstam, M. 1988. Terrorism on the high seas: the *Achille Lauro*, piracy and the IMO Convention on Maritime Safety. *American Journal of International Law* **82** (3), 269–310.

Harris, S. (ed.). 1984 *Australia's Antarctic policy options*. CRES Monograph II. Canberra: Australian National University.

Hassan, T. 1976. Third Law of the Sea Conference; fishing rights of land-locked States. *Lawyer of the Americas* **8** (3), 686–742.

Hodgson, R. D. & R. W. Smith. 1979. Boundary issues created by extended national marine jurisdiction. *Geographical Review* **69** (4) 423–33.

Howard, H. N. 1974. *Turkey, the Straits and US policy*. Baltimore: The Johns Hopkins University Press.

International Institute for Transportation and Ocean Policy Studies (until 1987 Dalhousie Ocean Studies Programme), Halifax, NS: Dalhousie University. Publishes studies on various aspects of the Law of the Sea and marine affairs.

Jados, S. S. 1975. *Consulate of the sea and related doocuments*. Ala.: The University of Alabama Press.

Janis, M. W. 1976. *Sea power and the Law of the Sea*. Lexington, Mass. and Toronto: D. C. Heath.

Jayaraman, K. 1982. *Legal regime of islands*. New Delhi: Marwah.

Jhabvala, F. (ed.) 1983. *Maritime issues in the Caribbean*. Miami: University Presses of Florida.

Johnston, D. M. 1987. *Canada and the new international law of the sea*. Toronto, Buffalo, London: University of Toronto Press.

Johnston, D. M., E. Gold & Phiphat Tangsubkul. (eds). 1983. *International symposium on the new Law of the Sea in Southeast Asia: developmental effects and regional approaches*. Halifax: Dalhousie Ocean Studies Programme.

Johnston, D. M. & P. M. Saunders. 1988. *Ocean boundary making; regional issues and developments*. London, New York and Sydney: Croom Helm.

Jonsson, H. 1982. *Friends in conflict – the Anglo-Icelandic cod wars and the Law of the Sea*. London: C. Hurst; Hamden, Connecticut: Archon Books.

Joseph, S. L. 1981. Legal issues confronting the exploitation of renewable resources of energy from the oceans. *California Western International Law Journal* **11**, 397–424.

Joyner, C. C. 1985. The Southern Ocean and marine pollution: problems and prospects. *Case Western Reserve Journal of International Law* **17** (2), 165–94.

Joyner, C. C. & S. K. Chopra (eds). 1988. *The Antarctic legal regime*. Dordrecht, Boston and London: Martinus Nijhoff.

Kent, G. 1980. *The politics of Pacific island fisheries*. Boulder, Colo.: Westview.

Kent, G. 1983. Regional approaches to meeting national marine interests. *Contemporary Southeast Asia* **5** (1), 86ff.

Kindt, J. W. 1984. Ocean thermal energy conversion. *Georgia Journal of International and Comparative Law* **14**, 1–27.

Knauss, J. A. 1984. The effects of the Law of the Sea on future marine scientific research and of marine scientific research on the future Law of the Sea. *Louisiana Law Review* **45**, 1201–20.

Koh, K. L. 1982. *Straits in international navigation; contemporary issues*. London, Rome and New York: Oceana.

Koh, T. T. B. 1987. The territorial sea, contiguous zone, straits and archipelagoes under the 1982 Convention on the Law of the Sea. *Malaya Law Review* **29**, 163–99.

Lapidoth, Ruth. 1983. The Strait of Tiran, the Gulf of Aqaba and the 1979 treaty of peace between Egypt and Israel. *American Journal of International Law* **77** (1), 84–108.

Lapodoth-Eschelbacher, R. 1982. *The Red Sea and the Gulf of Aden*. The Hague, Boston and London: Martinus Nijhoff.

Laursen, F. 1983. *Superpower at sea; US ocean policy*. New York: Praeger Scientific.

The Law of the Sea 1986. A series of articles first published in *Acta Juridica*. Cape Town: Juta.

The Law of the Sea Institute at the University of Hawaii publishes the proceedings of its annual conferences, occasional papers and workshop reports that are most valuable.

Lee Y. L. 1978. Malacca Strait, Kra Canal, and international navigation. *Pacific Viewpoint* **19** (1), 65–74.

Lee Y. L. 1980. *Southeast Asia and the Law of the Sea*. Singapore: Singapore University Press.

Legault, L. H. & B. Hankey. 1985. From sea to seabed: the single maritime boundary in the Gulf of Maine case. *American Journal of International Law* **79** (4), 961–91.

Leifer, M. 1978. *Malacca, Singapore and Indonesia*. Alphen aan den Rijn, The Netherlands: Sijthoff and Noordhoff.

Leiner, F. C. 1983. Maritime security zones: prohibited yet perpetuated. *Virginia Journal of International Law* **24**, 967–92.

Lien, J. & R. Graham (eds). 1985. *Marine parks and conservation* Toronto: National and Provincial Parks Association of Canada.

Limits in the seas. Authoritative monographs on marine boundaries and limits around the world (over 100 to date), issued periodically by the Bureau of Oceans and International Environmental and Scientific Affairs, Department of State, Washington, DC.

Lovering, J. F. & J. R. V. Prescott. 1979. *Last of lands. . . . Antarctica.* Melbourne: Melbourne Univesity Press.

Lumb, R. D. 1978. *The Law of the Sea and Australian off-shore areas.* Queensland: University of Queensland Press.

MacDonald, C. G. 1980. *Iran, Saudi Arabia, and the Law of the Sea; political interaction and legal development in the Persian Gulf.* Westport, Conn.; London, England: Greenwood Press.

Mangone, G. J. 1988. *Marine policy for America.* Lexington, Mass. and Toronto: Heath.

Mankabady, S. 1984. *The international maritime organization.* London and Sydney: Croom Helm.

Marine Policy. London: Butterworths. Published quarterly since 1976.

Marine Policy Reports. Monographs published 4 to 6 times a year by the Center for the Study of Marine Policy, University of Delaware 1977–89; quarterly journal since 1989.

Maritime Policy and Management. London: Taylor and Francis. Published quarterly since 1973.

Masterson, W. E. 1970. *Jurisdiction in marginal seas: with special reference to smuggling.* Port Washington, New York and London, England: Kennikat Press. (First published in 1929.)

McDorman, T. L., K. P. Beauchamp & D. M. Johnston. 1983. *Maritime boundary delimitation: an annotated bibliography.* Lexington, Massachusetts and Toronto: Heath.

McDougal, M. S. & W. T. Burke. 1987. *The public order of the oceans: a contemporary international Law of the Sea.* Dordrecht: Martinus Nijhoff.

M'Gonigle, R. M. & M. W. Zacher. 1979. *Pollution, politics, and international law; tankers at sea.* Berkeley, Los Angeles, London: University of California Press.

Miles, E. L., S. Gibbs, D. Fluharty, C. Dawson & D. Teeter. 1982. *The management of marine regions: the North Pacific.* Berkeley: University of California Press.

Miles, E. L. et al. 1985. *Nuclear waste disposal under the seabed.* Policy Papers in International Affairs no. 22. Institute of International Studies, University of California at Berkeley.

Mirvahabi, F. 1979. The rights of the landlocked and geographically disadvantaged states in exploitation of marine fisheries. *Netherlands International Law Review* **26** (2), 130–62.

Mitchell, B. & R. Sandbrook. 1980. *The management of the Southern Ocean.* London: International Institute for Environment and Development.

Molde, J. 1982. The status of ice in international law. *Nordisk Tidsskrift for International Ret.* **51** (3–4), 164–78.

Morris, M. A. 1987. *Expansion of Third World navies.* London: Macmillan.

Morris, M. A. (ed.). 1988. *North-south perspectives on marine policy.* Boulder and London: Westview.

Naghmi, S. H. 1980. Exclusive economic zone and the landlocked states. *Pakistan Horizon* **33** (1 & 2), 37–48.

Nash, M. L. 1981. US maritime boundaries with Mexico, Cuba and Venezuela. *American Journal of International Law* **75** (1), 161–2.

Occasional Papers/Reprints Series in Contemporary Asian Studies. Baltimore: University of Maryland Law School. A number of titles in this series deal with various aspects of marine affairs.

Ocean Development and International Law. New York: Taylor and Francis. Published quarterly since 1973.

Ocean and Shoreline Management. Amsterdam: Elsevier Scientific. Published quarterly since 1974.

Ocean Yearbook. Chicago: University of Chicago Press and Msida: International Ocean Institute of Malta. Published since 1978.

Oceanus; The International Magazine of Marine Science and Policy. Woods Hole, Mass.: Woods Hole Oceanographic Institution. Published quarterly since 1957.

O'Connell, D. P. 1982, 1984. *The international Law of the Sea*. vols 1 & 2. New York: Clarendon Press.

Oda, S. 1983. Fisheries under the United Nations Convention on the Law of the Sea. *American Journal of International Law* 77 (4), 739–55.

Orrego V., F. (ed.) 1983. *Antarctic resources policy; scientific, legal and political issues*. Cambridge: Cambridge University Press.

Orrego V., F (ed.). 1984. *The exclusive economic zone; a Latin American perspective*. Boulder, Colo.: Westview.

Papadakis, N. 1977. *The international legal regime of artificial islands*. Leyden, Netherlands: Sijthoff.

Papadakis, N. & M. Glassner (eds.) 1984. *The international Law of the Sea and marine affairs: a bibliography*. The Hague: Martinus Nijhoff.

Park, C. 1983. *East Asia and the Law of the Sea*. Seoul: Seoul National University Press.

Payne, R. J. 1979. Southern Africa and the law of the Sea: economic and political implications. *Journal of Southern African Affairs* 4 (2), 175–86.

Pharand, D. 1988. *Canada's Arctic waters in international law*. Cambridge: Cambridge University Press.

Pharand, D. & L. H. Legault. 1984. *The Northwest Passage; Arctic straits*. Dordrecht, Boston and Lancaster: Martinus Nijhoff.

Platzöder, R. (ed.). (1982 et seq.) *Third United Nations Conference on the Law of the Sea*: documents. Dobbs Ferry, New York: Oceana.

Polar Research Board, National Research Council. 1986. *Antarctic Treaty System*. Washington, DC: National Academy Press.

Polar Times. Rego Park, NY: American Polar Society. Published semiannually since 1935.

Pontecorvo, G. 1986. *The new order of the oceans; the advent of a managed environment*. New York: Columbia University Press.

Post, A. M. 1983. *Deepsea mining and the Law of the Sea*. The Hague, Boston and Lancaster: Martinus Nijhoff.

Pounds, N. J. G. 1959. A free and secure access to the sea. *Annals of the Association of American Geographers* 49 (3), 256–68.

Prescott, J. R. V. 1975. *The political geography of the oceans*. Vancouver and London: David and Charles.

Prescott, J. R. V. 1981. Australia's maritime claims and the Great Barrier Reef. *Australian Geographical Studies* 19 (1), 99–106.

Prescott, J. R. V. 1985. *Australia's maritime boundaries*. Canberra: Department of International Relations, The Australian National University.

Prescott, J. R. V. 1985. *The maritime political boundaries of the world*. London and New York: Methuen.

Puri, R. 1989. Exclusive political zone: a new dimension in the Law of the Sea. *Indian Political Science Review*, 14 (1), 39–54.

Quartermain, L. B. 1971. *New Zealand and the Antarctic*. Wellington: New Zealand Government Printers.

Ramazani, R. K. 1979. *The Persian Gulf and the Strait of Hormuz*. Alphen aan den Rijn, The Netherlands: Sijthoff and Noordhoff.

Ranken, M. B. F. (ed.) 1984. *Britain and the sea; future dependence–future opportunities*. Edinburgh: Scottish Academic Press.

Rao, P. C. 1983. *The new law of maritime zones*. New Delhi: Milind.

Rao, P. S. 1975. *The public order of ocean resources; a critique of the contemporary Law of the Sea*. Cambridge, Mass.: Massachusetts Institute of Technology.

Rembe, N. S. 1980. *Africa and the international Law of the Sea*. Alphen aan den Rijn, The Netherlands; Germantown, Maryland: Sijthoff & Noordhoff.

Rhee, S. 1988. Sea boundary delimitations between states before World War II. *American Journal of International Law* **82** (3), 443–58.

Richardson, E. L. 1988. Jan Mayen in perspective. *American Journal of International Law* **82** (3), 443–58.

Richardson, J. G. 1985. *Managing the ocean; resources, research law*. Mt Airy, Md: Lomond Publications.

Robertson, H. B., Jr. 1984. Navigation in the exclusive economic zone. *Virginia Journal of International Law* **24** (4), 865–915.

Rodgers, P. E. J. 1981. *Midocean archipelagos and international law*. New York: Vantage Press.

Ross, D. A. 1980. *Opportunities and uses of the ocean*. New York, Heidelberg and Berlin: Springer-Verlag.

Rothwell, D. R. 1988. *Maritime boundaries and resource development: options for the Beaufort Sea*. Calgary: the Canadian Institute of Resources Law, University of Calgary.

Rozakis, C. L. & C. A. Stephanou. 1983. *The new Law of the Sea*. Amsterdam, New York and Oxford: North-Holland.

Rozakis, C. L. & P. N. Stagos. 1987. *The Turkish straits*. Dordrecht, Boston and Lancaster: Martinus Nijhoff.

Samuel, M. S. 1982. *Contest for the South China Sea*. New York and London: Methuen.

San Diego Law Review Annual issue on the Law of the Sea.

Sanger, C. 1987. *Ordering the oceans: the making of the Law of the Sea*. Toronto and Buffalo: University of Toronto Press.

Schmidhauser, J. R. & G. D. Totten III (eds). 1978. *The whaling issue in US-Japan relations*. Boulder, Colo.: Westview.

Sebenius, J. K. 1984. *Negotiating the Law of the Sea*. Cambridge, Mass. and London: Harvard University Press.

Shapley, D. 1985. *The seventh continent; Antarctica in a resource age*. Washington, DC: Resources for the Future.

Sherman, K. & L. M. Alexander (eds). 1986. *Variability and management of large marine ecosystems*. Boulder, Colo.: Westview Press for American Association for the Advancement of Science (AAAS).

Singh, N. 1978. *Maritime flag and international law*. Leyden, The Netherlands: A. W. Sijthoff.

Smith, B. 1982. Innocent passage as a rule of decision: navigation v. environmental protection. *Columbia Journal of Transnational Law* **2**, 49–102.

Smith, G. P., II. 1980. *Restricting the concept of free seas; modern maritime law re-evaluated*. Huntington, NY: Robert E. Krieger.

Smith, H. D. 1984. The role of the sea in the political geography of Scotland. *Scottish Geographical Magazine* **100** (3), 136–50.

Smith, R. W. 1986. *Exclusive economic zone claims: an analysis and primary documents*. Dordrecht, Boston and Lancaster: Martinus Nijhoff.

Song, Y. 1988. The British 150-mile fishery conservation and management zone around the Falkland (Malvinas) Islands. *Political Geography Quarterly* **7** (2), 183–97.

Soons, A. H. A. 1982. *Marine scientific research and the Law of the Sea.* Deventer, Netherlands: Kluwer Law and Taxation Publishers.

Soule, D. F. & D. Walsh (eds). 1983. *Waste disposal in the oceans.* Boulder, Colo.: Westview.

Stone, J. C. (ed.) 1985. *Africa and the sea.* Aberdeen: University African Studies Group.

Sulaiman, M. A. 1984. Free access; the problem of land-locked States and the 1982 United Nations Convention on the Law of the Sea. *South African Yearbook of International Law* **10**, 144–61.

Symmons, C. R. 1979. *The maritime zones of islands in international law.* The Hague, Boston and London: Martinus Nijhoff.

Symmons, C. R. 1986. Maritime boundary disputes in the Irish Sea and northeast Atlantic Ocean. *Marine Policy Reports* pp. 2, 4, 5.

Symonides, J. 1988. *The new Law of the Sea.* Warsaw: Polish Institute of International Affairs.

Symposium: The International Legal Regime for Antarctica. 1986. *Cornell International Law Journal* **19** (2).

Taitt, B. M. 1983–4. The exclusive economic zone: a Caribbean community perspective. *West Indian Law Journal,* Part I, vol. 7, pp. 36–55; Part II, Vol. 8, pp. 26–44.

Tangsubkul, P. 1982. *ASEAN and the Law of the Sea.* Singapore: Institute of Southeast Asian Studies.

Tangsubkul, P. 1984. *The Southeast Asian archipelagic states: concept, evolution and current practice.* Honolulu: East-West Center.

Theutenberg, B. J. 1983. "The Arctic Law of the Sea". *Nordisk Tidsskrift for International Ret* **52** (1–2), 3–39.

Theutenberg, B. J. 1984. *The evolution of the Law of the Sea. A study of resources and strategy with special regard to the polar areas.* Dublin: Tycooly International.

Torres, N. T. 1985. Annotated bibliography: delimitation of exclusive economic zone boundaries between opposite and adjacent states and the Gulf of Maine dispute. *Maryland Journal of International Law and Trade* **9**, 181–6.

Treves, T. 1980. Military installations, structures and devices on the seabed. *American Journal of International Law* **74** (4), 808–57.

Triggs, G. D. (ed.). 1987. *The Antarctic treaty regime; law, environment and resources.* Cambridge: Cambridge University Press.

Truver, S. C. 1980. *The Strait of Gibraltar and the Mediterranean.* Alphen aan den Rijn, The Netherlands: Sijthoff and Noordhoff.

Underdal, A. 1980. *The politics of international fisheries management – the case of the northeast Atlantic.* Oslo: Universitetsforlaget.

United Nations 1989. *The Law of the Sea; baselines: an examination of the relevant provisions of the United Nations Convention on the Law of the Sea.* Office for Ocean Affairs and the Law of the Sea. Sales. E.88.V.5.

United Nations. *The new Law of the Sea Bulletin.* Issued irregularly by the Law of the Sea Secretariat since 1983.

United Nations. *The Law of the Sea; national legislation on the exclusive economic zone, the economic zone and the exclusive fishery zone.* Sales no. E.85.V.10

United Nations. *The Law of the Sea; multilateral conventions relevant to the United Nations Convention on the Law of the Sea.* Sales no. E.85.V.11.

United Nations. *The Law of the Sea; rights of access of land-locked States to and from the sea and freedom of transit.* Sales no. E.87.V.5.

United Nations Environment Programme 1985. *Environment and resources in the Pacific*. Nairobi: UNEP.

United Nations Environment Programme 1988. *Medwaves*, no. 12.

Valencia, M. J. 1980. The South China Sea: constraints to marine regionalism. *Indonesian Quarterly* **8** (2), 16–38.

Valencia, M. J. (ed.). 1981. *The South China Sea: hydrocarbon potential and possibilites of joint development*. Oxford: Pergamon. (Special issue of *Energy* **6**, 1.)

VanderZwaag, D. L. 1983. *The fish feud*. Lexington, Mass.: D.C. Heath.

Van Dyke, J. & S. Heftel. 1981. Tuna management in the Pacific: an analysis of the South Pacific forum fisheries agency. *University of Hawaii Law Review* **3** (1), 1–65.

Van Meurs, L. H. 1985. *Legal aspects of marine archaeological research*. Cape Town: Institute of Marine Law, University of Cape Town.

Vertzberger, Y. Y. I. 1984. *Coastal states, regional powers, superpowers and the Malacca-Singapore straits*. Berkeley: Institute of East Asian Studies, University of California.

Vukas, B. (ed.). 1988. *The legal regime of enclosed or semienclosed seas: the particular case of the Mediterranean*. Zagreb: Institute of International Law and International Relations, University of Zagreb.

Walton, K. 1974. A geographer's view of the sea. *Scottish Geographical Magazine*. **90** (1), 4–13.

Westerman, G. 1987. *The juridical bay*. New York: Oxford University Press.

Westermeyer, W. E. & K. M. Shusterich (eds). 1984. *United States Arctic interests; the 1980s and the 1990s*. New York: Springer-Verlag.

White, A. T. 1984. Marine parks and reserves: management for Philippine, Indonesian and Malaysian coastal reef environments. PhD dissertation in geography, University of Hawaii.

Wijkman P. M. 1982. UNCLOS and the redistribution of ocean wealth. *Journal of World Trade Law* **16** (1), 27–48.

Wiktor, C. . & L. A. Foster. 1987. *Marine affairs bibliography; a comprehensive index to marine law and policy literature; cumulation: 1980–1985*. Dordrecht, Boston, Lancaster: Martinus Nijhoff.

Wolfrum, R. & K. Bockslaff (eds). 1984. *Antarctic challenge; conflicting interests, cooperation, environmental protection, economic development*. Berlin: Duncker & Humblot.

Wooster, W. S. 1984. Sea law and ocean research: view from the northwest. *Oregon Law Review* **63**, 121–37.

Index